.

FLUID MOTION UNMASKED

Exploring the Dynamics of Flows
in Nature and Technology

FLUID MOTION UNMASKED

Exploring the Dynamics of Flows in Nature and Technology

Michael Leschziner

Imperial College London, UK

World Scientific

NEW JERSEY · LONDON · SINGAPORE · BEIJING · SHANGHAI · HONG KONG · TAIPEI · CHENNAI · TOKYO

Published by

World Scientific Publishing Europe Ltd.

57 Shelton Street, Covent Garden, London WC2H 9HE

Head office: 5 Toh Tuck Link, Singapore 596224

USA office: 27 Warren Street, Suite 401-402, Hackensack, NJ 07601

Library of Congress Cataloging-in-Publication Data

Names: Leschziner, Michael author
Title: Fluid motion unmasked : exploring the dynamics of flows in nature and technology /
    Michael Leschziner, Imperial College London, UK.
Description: London ; Hackensack, NJ : World Scientific Publishing Europe Ltd, [2026] |
    Includes bibliographical references and index. | Contents: Put your senses to work --
    The many faces of fluid mechanics. Pinnacle of complexity -- Back to basics --
    Life-support systems -- Whirling, ascending and mixing --Global dance --
    Closing remarks -- Unmasking the hidden faces and interior of flows.
    Observing and measuring -- Modelling and number-crunching.
Identifiers: LCCN 2025013287 | ISBN 9781800617582 hardcover |
    ISBN 9781800617742 paperback | ISBN 9781800617599 ebook |
    ISBN 9781800617605 ebook other
Subjects: LCSH: Fluid mechanics
Classification: LCC TA357 .L475 2026 | DDC 620.1/06--dc23/eng/20250618
LC record available at https://lccn.loc.gov/2025013287

British Library Cataloguing-in-Publication Data
A catalogue record for this book is available from the British Library.

For any available supplementary material, please visit
https://www.worldscientific.com/worldscibooks/10.1142/Q0518#t=suppl

Desk Editors: Murali Appadurai/Gabriel Rawlinson/Shi Ying Koe

Typeset by Stallion Press
Email: enquiries@stallionpress.com

To Maya, Ava, Sienna and Ethan, who may, one day, be tempted to dig into their grandfather's curious obsession.

Preface

Until my retirement, in 2023, I had been a university academic for 43 years, most recently as Professor of Computational Aerodynamics at Imperial College London. During that time, my life was consumed with the challenges of doing research at the leading edge of computational fluid dynamics, with the rigours of publishing my work in learned journals, with the race for citation counts and — critical to progression — with securing the best scores I could from students attending my lectures. As my research was fundamental in nature, mostly at a fair distance from real-world engineering application, I only rarely thought about what "Joe and Jane Public" — the ordinary, non-expert, men and women in the street — thought about the subject of fluid mechanics, in general, and about my focus on turbulence phenomena in fluid flows, in particular. However, one dark corner in my memory that has had an unsettling influence on my life in the comfortable academic echo chamber is due to my daughter, Tamara, now in her late 40s.

In the 1980s and 1990s, I was working at the University of Manchester Institute of Science and Technology (UMIST), doing research on turbulence modelling. My "office" — cum dining room — was then closest to the main door of our house, so I was the one who usually let Tamara's friends in when she had birthday parties. Ahead of these parties, Tamara gave me two strict instructions: "do not ask my friends … *and who are you?*" and "do not talk to my friends about fluid mechanics". While both demands were reasonable, I felt rather deflated by the second because its implicit message was that fluid mechanics was boring and "for the birds".

Thirty-five years have now passed, and I suspect my daughter's aversion to fluid mechanics has not changed greatly. However, I am pinning my hopes on my granddaughter Maya who is displaying, albeit so far only at high-school level, a keen interest in mathematics and science. It is primarily in response to her prompts — "grandad, why don't you write a fluid mechanics book for dummies" — that I decided to embark on writing this text. In fact, it is fair to say that it is Maya who is the imaginary target of this book; I dare not think about how many others would be interested in my musings. My idea was that if I wrote a text in accessible terms that are likely to be understood by an intelligent 16-year-old, the text might also appeal to some non-expert adults. I am not aware of any book of this kind in the open literature, despite the fact that the need to educate the wider public about fluid mechanics is well appreciated (see, for example, "Taking Fluid Mechanics to the General Public" by Etienne Guyon and Marie Yvonne Guyon, *Annual Reviews of Fluid Mechanics*, volume 46, 2014).

Readers might be surprised to learn that I found it very hard to write this book, and I may well not have succeeded in fully meeting Maya's or anybody else's expectations. First, it touches upon many topics that were unfamiliar to me. Fluid mechanics is a very broad subject with hundreds if not thousands of facets, directions, topics and applications. It pertains to numerous natural phenomena, processes and engineering equipment, having a huge range of scales, dimensions, appearances, features and operational objectives. For example, what connects a microscopic "Lab on a Chip", used for rapid medical diagnostics, to big ocean liners? What do krill swimming in oceans and coronal injections from the sun have in common? What connects blood flow in our cardio-vascular system with magma under the Earth's crust or with the Gulf Stream? Thus, unless one aims to write a multi-volume encyclopaedic text, it is unrealistic to try to cover the subject in anything like a comprehensive fashion. All that I try to do with this book is to argue why and how fluid mechanics affect every facet of our lives and to encourage readers to observe, think about and explore phenomena they have routinely passed by without much thought or attention before reading this book.

The second challenge in writing this book lies in the fact that fluid mechanics is a highly mathematical subject, involving many principles that can only be fully clarified with equations and differential calculus. Yet, I set out to write a text that is devoid of mathematics — or almost devoid of mathematics, as will transpire. Hence, I found myself often having to rely on intuitive arguments and woolly explanations that the purist might scoff at and regard as inadequate for lacking rigour and for not offering full transparency.

Since air, wind, water and rain are such fundamental ingredients of human life, rational observations of fluid mechanic phenomena probably go back to the beginning of humans' conscious awareness of the world around them. Our knowledge about this awareness relies on mankind's ability to record events on durable media.

Simple depictions of water waves go as far back as 3000 BC, to ancient Egypt, the hieroglyph for water being WWV.

Assyrian wall plate showing waves and eddies as part of a scene depicting a military battle close to a body of water, ca. 700 BC.

Source: © The British Museum (with permission).

Waves and whirls in rivers were described on carved tablets some 3000 years ago. For example, the plate above shows a close to 2700-year-old Assyrian relief of a military battle at a river or the sea in which different phenomena associated with waves and eddies are insightfully depicted. This is only one of several carved Assyrian recordings of scenes featuring river flows, waves and eddies, displayed in the British Museum.

Even older are Bronze-age cave drawings of boats apparently being upended by large turbulent eddies. Finally, the ancient Greeks, among several other ancient civilisations or religions, believed that the world was composed of 4–8 elements, all including air (wind), water and fire, the last being simply a hot, chemically reacting, turbulent gas flow of which air is the main constituent.

Three thousand years later, we are still scratching our heads about precisely what makes turbulent eddies tick, and we still struggle to understand how turbulent flows behave in detail. Yet, en route to today's modern world, we somehow managed to construct and operate nuclear power plants, build jet engines and rockets, formulate weather-prediction models and create aeroplanes that fly across 15000 km in less than a day, the function or operation of all of which hinges on turbulent flows.

One especially challenging aspect of fluid mechanics is that fluid flows are, with very few exceptions, "nonlinear" in character. The term "nonlinear" is not easy to describe in words because what is actually nonlinear is the set of equations describing the flows. In essence, what is implied by the equations is that the behaviour of a flow, in particular, its evolution in time, can change dramatically — *nonlinearly* — and in an unpredictable fashion in response to small perturbations around the position and time at which the flow is considered. For example, if a small float is placed on the surface of a water flow well upstream of an obstacle, the exact path of the float behind the obstacle cannot be predicted, because even tiny errors in the description of the flow at the time the float is placed on the water surface, or tiny disturbances at any stage at which the float moves downstream, can have profound consequences on the end point of the float. This difficulty is encapsulated in the term "The Butterfly Effect" in weather prediction: any insignificant disturbance anywhere on the globe has, ultimately, a difficult-to-predict effect on the weather at future times. The inevitable consequence of nonlinearity is, therefore, that any flow having non-trivial complexity can evolve into an innumerable number of states.

In the face of these characteristics, one might be justified in assuming that understanding and describing fluid flows are hopeless objectives. Fortunately, this is not so because scientists have evolved tools that allow a sufficient level of understanding, associated mathematical tools and computer algorithms to be derived that allow the natural environment to be analysed, at least approximately, and useful engineering solutions to be developed. The present book attempts to convey this fact.

One further comment is appropriate here in relation to nonlinearity. Since almost any flow is very sensitive to its environment, it follows that even slight changes in the geometry containing the flow can have profound effects on the flow and thus the operational characteristics of the associated device exploiting the flow properties. This sensitivity not only presents challenges but also offers opportunities. Thus, we can develop a range of engineering solutions by introducing relatively slight modifications to a basic type of geometry. For example, we can change the ventilation characteristics in a room substantially by opening or closing a small window at a well-chosen strategic location in the room. Similarly, we can substantially change the operation of a combustor by slightly sloping the combustor walls or introducing a small step or obstacle at a well-chosen location. This enormous variability is yet another reason why it is not realistic to try to be inclusive in any discussion or description of fluid-flow phenomena.

My hope is that the description contained in this book, however incomplete and selective, will suffice to convey the message that fluid flows are not merely fascinating but critical to our existence. They thus deserve to be understood and appreciated by all, not merely by relatively few physicists who strive for insight and engineers who exploit the flow physics to develop new life-enhancing processes and equipment.

About the Author

Michael Leschziner is Emeritus Professor of Computational Aerodynamics in the Aeronautics Department at Imperial College London and is Fellow of the Royal Academy of Engineering. He specialises in the physics and modelling of turbulence and its representation in computational algorithms applied to fluid flows in mechanical and aeronautical engineering. He was previously Head of the Thermo-Fluids Division and of the Mechanical Engineering Department at the University of Manchester Institute of Science and Technology and Professor of Aerospace Engineering at Queen Mary University of London. He has published numerous articles in the academic literature, and his previous book, titled *Statistical Turbulence Modelling for Fluid Dynamics — Demystified*, was published in 2015 by Imperial College Press.

Contents

Chapter 1

Put Your Senses to Work

Take a short break from your frantic daily routine and make a conscious effort to look at, feel and listen to the *life of fluid flows* around you, which most of us take for granted without much thought or question...

··· Look out of your window and observe the seemingly chaotic motion of clouds, tree branches, leaves and bushes or clothing hanging on a washing line on a windy day. What is the origin of this apparent randomness?

··· With a little more patience, you might see a bird or a bee flapping its wings to remain aloft and land elegantly on the ground, a tree branch or a flower. How do these creatures stay aloft and control their flight so gracefully?

··· Open a tap in your kitchen or bathroom and feel and listen to the water as it streams through your fingers and washes down the plug hole. Ask yourself why the surface is smooth (like Golden Syrup) in very weak flow but turns rough and wavy as you open the tap further.

··· Observe the flow pattern on the bottom of the kitchen sink as the water jet hits the surface and spreads out, forming a distinct circular rim at which the water level rises sharply and the radial speed declines suddenly. What is the origin of this stepwise change?

··· Light a match or a candle and observe the motion of the smoke rising in an intricate combination of meandering vortices and swirls; observe the intricately swirling smoke or steam discharged by a chimney on a cold day, eventually dissipating as it is "absorbed" by the

1

atmospheric air. Why are these flows so chaotic rather than steady and ordered?

··· Listen to the whistling, whining and buzzing sounds and vibrations provoked by a stormy wind as it flows past trees, houses, electricity poles and wires. What causes this high-pitch sound?

··· Feel the stream of air flowing through your fingers as you extend your hand out a window of a car driven at high speed, and question what precisely causes the resistance on your hand, or that opposing you when riding a pushbike, or inverting your umbrella in windy conditions.

··· On your next train journey or trip on the Underground (Metro), note the air blast hitting you as the nose of the train passes by and the wind-like slipstream of air that is flowing along the platform as the train slows down. What process drags the air along the train?

··· Observe how dried leaves and bits of paper accumulate in corners and behind buildings, performing a circular dance along the ground in windy conditions. What mechanism causes this odd behaviour?

··· Look at the rising bubbles in a cooking pan in which water is being brought to a boil on a hot plate and the steam rising and dissipating above the surface. Get closer to the water surface with the palm of your hand and feel the steam and air heating your skin. Why is the temperature of the steam reaching your hand so much lower than that of the boiling water?

··· Listen to your breathing and the whooshing sound that is produced when you blow air out at high speed through your pursed lips; observe a high-speed jet and spray exiting from a garden hose connected to a nozzle gun, and note the noise emanating from the fast stream of water; listen to the intense roar and hiss emitted from the jet engines of a passenger airliner taking off or landing. What processes connect the three seemingly very different flows, and what is the origin of the noise?

··· Listen to the machine-gun-like popping sound created by a helicopter flying overhead. What is the cause of this noise?

··· Place yourself in front of a fan and feel the intermittency of the cooling air when the fan is turned at a slight angle from your face. What is the origin of this pulsation at the edge of the stream?

··· Observe the manner in which long hair executes a wavey, swirling dance when subjected to the fast stream of air emitted by a hair drier. Why is this motion so unsteady?

··· Feel your racing pulse, when exercising or running, driven by the heart pumping blood through your arteries. How do the lungs and heart collaborate to provide just the right amount of oxygen and blood to sustain your exertions?

··· Observe the propagation of waves on any calm water surface and listen to the noise emanating from the waves breaking on a shore. What are the mechanisms by which the shallow waves rise up and break as they approach land?

··· Focus on a saxophone or clarinet being played and contemplate how the air flow interacts with the reed to cause the air to vibrate and produce the pleasing sound.

For every one of these many manifestations of fluid flow, there are innumerable others, many far less obvious and not easily accessible to our senses. Many are facets of our natural world — for example, the flow of blood and other liquids through various organs in our body, underground water streams and magma motion below the Earth's crust, oceanic circulation, large-scale weather systems, solar storms and plasma flows within stars. Many more flows are of critical importance to numerous technological applications — just think of flight, road transport, propulsion, energy conversion, geothermal power generation, oil refineries, liquid-food processing, wind, steam and gas turbines, pumping, heating, cooling, ventilation, pipe networks transporting water, oil and steam, river and flood management and sewage-treatment plants.

Why would we want to delve into the inner mechanics of the many flows within which we are emersed and the somewhat fewer ones which are within us? Intellectual curiosity aside, the answer is that, without understanding the mechanisms and processes that are driving these flows, we cannot hope to optimally harness, manipulate and control them for the purpose of protecting our natural environment, of improving our physical health by promoting fluids-based medical treatments, of reducing energy losses and thus improving the efficiency of the numerous technological applications on which we depend, and of mitigating the environmental damage that these applications cause.

Perhaps the biggest challenge of our era is how to halt, or at least slow down, global warming, while maintaining the integrity of our industrial infrastructure. We *sense* global warming through having to endure increasingly extreme weather conditions — heat waves, droughts, storms, prolonged periods of heavy rain and floods. Yet, we are not prepared to compromise on our technology-driven lifestyle. Our ability to resolve this dilemma depends on our understanding and, ultimately, influencing favourably the interactions among the atmosphere, the oceans and fresh-water resources by mitigating the detrimental consequences of techno-logical developments or replacing harmful technologies. Specifically, efforts to replace traditional combustion-based power generation and transport with environmentally sustainable methods, such as wind tur-bines and electrical propulsion, are central to addressing climate change. Alongside this realignment in the power-generation sector, the manipula-tion and control of fluid flows is another important route to reducing the detrimental impact of industrial flows through targeting energy losses. By "loss" we mean here the transfer of useful mechanical energy in a flow to useless heat. Such a loss arises because of a combination of turbulence and viscous friction within the flow. Turbulence, the subject of Chapter 5, is a state of extreme fluid agitation and disorder in a flow and a major source of loss of useful energy. Turbulence has a major influence on dis-persion and the manner in which flows interact with solid bodies. Understanding these interactions is a prerequisite for developing creative flow-control strategies.

In each and every natural and technological system or application, a distinct set of physical properties and characteristics of the fluid(s) in question are exercised or are being exploited. While every one of us will have developed a degree of intuitive "understanding" of how fluids behave in ordinary everyday situations, such as in some of the examples listed above, this understanding — being based on *seeing, feeling* and *hearing* — is highly superficial and lacks the insight that can only be acquired through a scientific description of the behaviour of fluid flows. This description is part and parcel of a subject of Physics referred to as "Fluid Mechanics". This subject is covered by a large body of published literature — many books and thousands of journal papers — reflecting the fact that the behaviour of fluids has been actively researched for, literally, hundreds of years.

Leonardo da Vinci (1452–1519) was, arguably, the first serious fluid dynamicist who developed an astonishing depth of understanding based on observation — *seeing* — and imaginative interpretation of many fluid-flow phenomena, as is exemplified in Figure 1.1. The subject, in its modern form, involves complex mathematics, advanced experimental techniques, numerical analysis and computer codes. These are employed by many advanced industrial laboratories to design fluid-based machinery

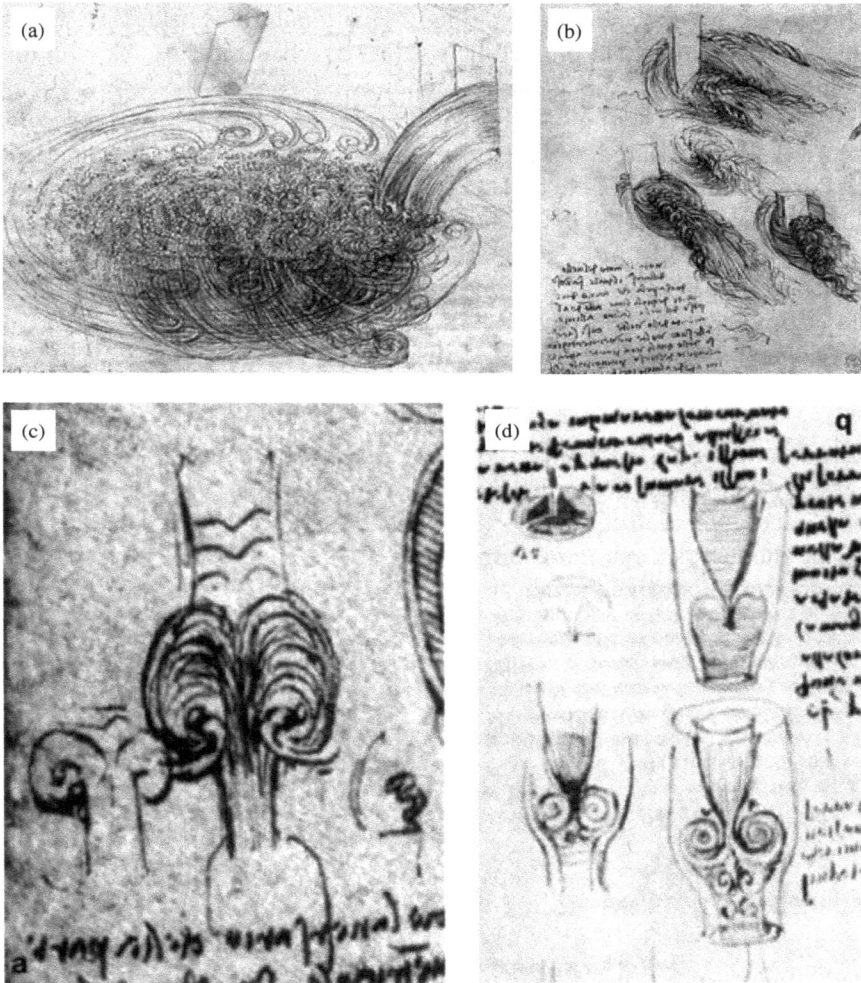

Figure 1.1.   Detail (part) images from four of the many drawings of fluid flows produced by Leonardo da Vinci (ca. 1470): (a) vortical structure of waste water flowing into a pool; (b) pattern of flows around and behind pillars and obstacles; (c,d) recirculating flow in an aortic enlargement.

Source: Royal Collection Enterprises Limited 2024 (with permission), © Royal Collection Trust.

and processes, and they are exploited by national meteorological offices and marine institutes to predict the weather and oceanic currents with the help of extremely powerful supercomputers.

Do the implications of the last two sentences above cause you concern? If so, be assured that this text will not delve into the mathematical analysis of fluid flows, beyond a highly rudimentary level, however helpful such an analysis would be in conveying a full understanding of the underlying physical processes. Rather, the aim is for the reader to gain an enhanced appreciation of the physical behaviour of fluids discussed in a descriptive manner that is accessible to non-specialists and is presented in the context of specific real-world situations. If the descriptive coverage to follow succeeds in conveying a greater degree of insight into the physical interactions which give rise to phenomena that can be sensed — *seen, heard, felt* — then the book will have served a useful purpose. At the very least, readers may hopefully develop a greater degree of appreciation of — and open their minds to — the hugely important role of fluid flows in the numerous environmental, biological and engineering systems that are critical to our lives.

Chapter 2

The Many Faces of Fluid Mechanics

2.1 Pinnacle of Complexity

We start, perhaps counterintuitively, at the complex end of the subject: rocket science, a truly mind-blowing technological application of fluid mechanics. We do so to establish an "upper norm" against which to contrast the many flows to follow and to illustrate the wide range of physical processes that have to work together to achieve a desired operational performance in an extremely elaborate fluid-flow scenario. There are several other candidates for the top complexity slot. A large jet engine is one; a nuclear power plant is another, both discussed later. We choose a rocket engine for its wholly exceptional power output — or power density — and its popular exposure in facilitating space exploration.

There can be no greater awe-inspiring experience than witnessing the ear-shattering, fire-spewing take-off of a large space rocket, such as the Saturn V shown in Figure 2.1. The sensory experience aside, the statistics are truly staggering:

- The first stage, incorporating five separate engines, produces 3500 tons of thrust, capable of lifting 10 fully laden and fully fuelled jumbo jets vertically upwards. In comparison, the largest jet engines powering civilian airliners produce 35 tons of thrust, relative to 700 tons produced by each Saturn V engine.
- The five engines consume around 13 tons of kerosene and liquid oxygen per second — yes, per second!

Figure 2.1.   The Saturn V rocket in ascent and one of its five stage-1 engines. The structure above the nozzle contains the fuel pumps.

Source: NASA.

- The power output of the rocket is around 120 GigaWatts, equivalent to 160 million horsepower and corresponding to the output of 10 large power stations, sufficient to power 50–100 million homes.
- The fuel and oxidant pumps in any one engine are rated at around 55000 horsepower, similar to the power output of a large civil aircraft jet engine operating in full thrust.
- The exhaust jet, issuing from the engine nozzle having a diameter of almost 4 meters, has a speed of around 2500–3000 meters per second, about 8 times the speed of sound, and its temperature is around 3000°C.

There are very few complex physical processes that do not feature in a rocket-engine flow: insanely fast, powerful, hot, burning, containing many chemical species, droplets and particles. Since the jet is supersonic, the existing mass "hits" the atmosphere, like a "brick wall" (admittedly, an amateurish description), with a speed that does not allow the air outside the jet to sense the pressure of the oncoming stream. As seen in the laboratory test in Figure 2.2, this causes a series of shock waves in the jet as it slows down — a sonic boom or explosive wave being a familiar relative. The exhaust gas thus "relaxes" to the pressure of the surrounding air, adapting itself to the outer environment, in a complex series of

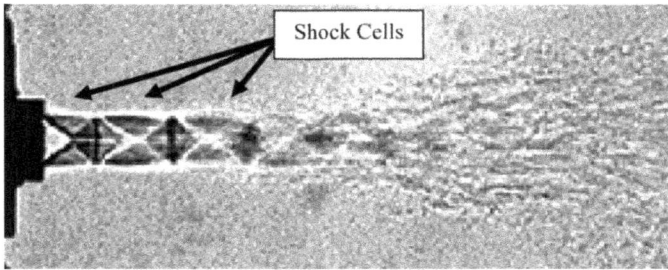

Figure 2.2. Laboratory visualisation of the shock structure in a highly supersonic non-reacting air jet, similar to that of a rocket-exhaust jet discharging into stagnant air.

Source: Liu *et al.* [1], Visualization Society of Japan (with permission).

compression and expansion waves in which the pressure, temperature and density vary over huge ranges.

There are countless technological applications of fluid flow that are much less specular than the Saturn V, but many are more important to us in our daily lives, even though we take them for granted. It is impossible to list more than a few: aircraft, ships, cars, petrol and diesel engines, jet engines, wind turbines, steam turbines, compressors, cooling fans, heaters, burners, pumps, gas boilers and hobs, heat pumps, heat exchangers, spray cans, medical inhalers, heart valves, heart-lung machines, dialysis machines and so it goes on as far as our mind's eye can take us. Many involve complex phenomena and interactions that are also operative in rocket engines — e.g., heat transfer, chemical reactions, species transport, bubbles, droplets and solid particles. A significant number of the technological fluid flows listed above are discussed in chapters to follow, especially in Chapter 6. However, with a few exceptions aside, we confine ourselves to "conventional" fluid mechanics, avoiding complications that arise from particles, bubbles and combustion chemistry, which tend to obscure the main fluid-mechanic phenomena on which this book intentionally focuses.

The natural-environment equivalent to the rocket engine, as an extreme-complexity system, is the global atmosphere and the oceanic system closely interacting with it. This is one subject considered in more detail in Sections 2.5 and 4.2, with particular emphasis on atmospheric flow phenomena.

At even larger (indeed, much larger) scale are planetary, stellar and cosmological flows, such as those shown in Figure 2.3. Many are unimaginably exotic in terms of their size, speed, temperature, energy

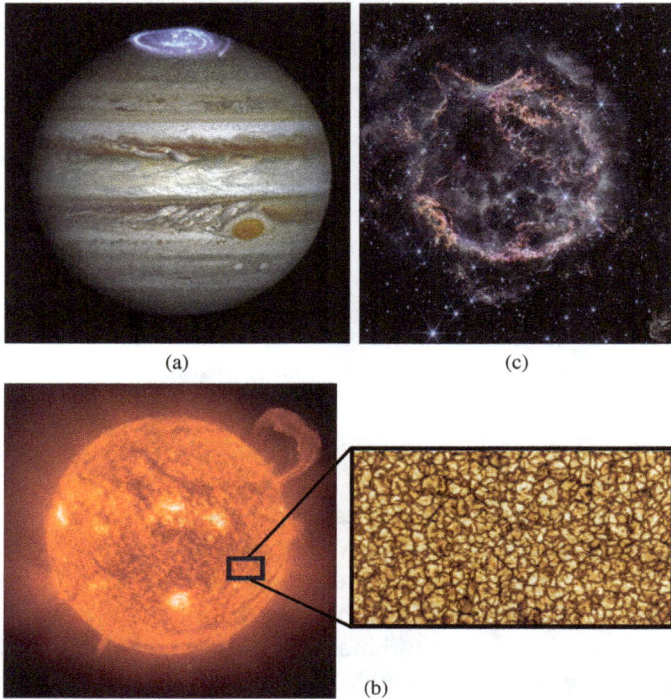

Figure 2.3.   Astronomical and cosmological flows: (a) a view of Jupiter with "Great Red Spot" and turbulent wake, taken by Juno spacecraft. The brightest features are at the highest altitudes — the tops of convective storm clouds; (b) highly buoyant, turbulent flow on the Sun's surface, including a large solar flare. The insert shows structures as small as 18 miles in size, a pattern of turbulent, "boiling" gas that covers the entire solar surface; (c) the expanding shells of debris from an exploded star — a "supernova" — taken by the James Web Telescope. The main ring features strong turbulence and is about 15 light-years across.

Source: NASA/JPL; SOHO — a NASA and ESA consortium.

output and sheer remoteness from our experience with terrestrial fluid flows.

Jupiter's atmosphere is characterised by violent storms of hydrogen and helium gases at a density almost 10 times lower than that of our ground-level atmosphere. The so-called "Great Red Spot", seen in Figure 2.3(a), is a persistent anti-cyclonic storm system of size roughly that of the Earth's diameter, in which the gas races at speeds around twice that of a category-5 hurricane (some astonishingly detailed images taken by NASA's Juno spacecraft can be found in https://www.jpl.nasa.gov/images/pia01353-jupiter/).

The Sun's surface, shown in Figure 2.3(b), appears as a highly turbulent collection of bubbling hot eddies of hydrogen and helium at a temperature of around 5500°C. The regular ejection of extremely powerful solar coronal flares often features in the news media for being harbingers of intense Northern Lights and disturbances in terrestrial electronic communications. The structure seen in the top right-hand corner of Figure 2.3(b) has a size roughly 30 times the Earth's diameter, the mass it contains is around 1 billion tonnes and its speed is of order 1000 km per second.

As staggering as the above solar-flow numbers may be, those pertaining to large cosmological gas flows are truly mind-boggling. The ionised gas shell resulting from a supernova, such as that shown in Figure 2.3(c), has a diameter of 10–15 light years. The outer surface of the shell identifies a shock wave that arises from the collision of the ionised gas ejected by the supernova explosion with cool interstellar gas. Although the density of the shell is extremely low — many billions of times lower than the density of our atmosphere — the processes within the shell involve scales that are beyond our comprehension. Thus, the shock moves outwards at a speed as high as 30000 km per second — roughly 10% of the speed of light — and the collision causes the temperature of the shell to rise to several millions °C.

Remarkably, in view of the huge disparity of its scales relative to terrestrial flows, the supernova gas shell has chaotic features akin to conventional turbulent flows, as will emerge from Chapter 5 which focuses on the subject of turbulence and its manifestations.

2.2 Back to Basics

We take a massive step back from the complex end of fluid mechanics, for a while, to consider some fundamental properties of fluid flows.

It is axiomatic that fluids move freely — or almost feely. Even a small child will accept and welcome without question the breeze that cools its face on a hot day, will delight in the splashing of water in a pool or a bath, will watch with interest the jet of water streaming out of a tap into a glass and will be displeased about the rain wetting its face and hands on a cold windy day.

In the case of a stream of water interfacing with the atmosphere or colour-infused liquids, we see the motion, we hear the flow, and we feel its force on our limbs and its temperature on our skin. This multi-sensory contact will evoke a comforting sense of familiarity with the phenomena, even if not conferring intellectual insight into the physical mechanisms involved. With air or gases, however, we rely on indirect observations: the rapid rustling of leaves and the slow rocking of branches in the wind, the waves driven by wind, smoke issuing from a chimney or produced by a fire, the clouds moving high above us by unseen atmospheric streams, the swirl of snowflakes, tiles being torn off roofs and fences being torn down by hurricane-force winds, and dust or sand being carried by gusts and rotating vortices.

All these observations may be qualitatively instructive, but what do we actually know about the mechanisms that make fluid flows tick? The answer is "very little" in most cases, simply because we do not understand the internal dynamics and the physics involved. In some respects, this is akin to our observing the motion of celestial bodies — planets, moons, stars and galaxies — superficially familiar, yet intellectually elusive, unless we delve seriously into astronomy and cosmology.

The ability of fluids to deform and move freely, in most cases with little resistance, is not merely a random choice made by nature but is an extremely important — indeed, existential — property. We consider more carefully later what is meant by "move freely" — just think about the resistance offered by viscous honey to appreciate that "freely" is a relative term. Here, we consider the rather unpleasant implications and consequences of trying to live in a virtual world in which fluids do not move freely, contrary to reality. More precisely, we pose the following question: what happens if fluids were stagnant and deprived of the ability to mix?

Imagine being immersed in an atmosphere that has the same physical properties (gaseous composition, viscosity, thermal conductivity and density) as real air, that can be freely displaced locally — say, by breathing in and out and by movement — but that is otherwise stagnant. This would be akin to being surrounded by a very fluffy gel-like substance. As you breathe in and out, you inhale and exhale one and the same mass — a "bubble of air" — as there is no mixing with the surrounding air between exhaling and inhaling. An obvious consequence is that you would perish within a few minutes as the oxygen is consumed and the CO_2 content builds up.

One other impossible challenge you would face is controlling your body temperature. The heat you release through your skin would heat the layer next to it, and its temperature would rise continuously. Some of the heat would diffuse outwards, by conduction, but this is a very weak process: air is an excellent insulator and its thermal conductivity is thousands of times lower than that of metals (10000 times in the case of copper). With the surrounding air temperature rising, your body temperature would also rise without control, and you would perish. The same would apply to any land-based animal. In fact, most vegetation would also very quickly die out as the exchange between CO_2 and O_2, on which photosynthesis depends, would be prevented. The only way to survive in this stagnant atmosphere — at least for a while — would be by moving through it constantly, thus changing the gaseous environment around you. Fish and sharks do this, but they, too, would face some serious limitations that would arise from the absence of mixing in their aquatic environment, which secures the supply of oxygen from the atmosphere into the water below.

So, what is it that prevents the hellish, lifeless world described above? The answer is *mixing by fluid motion*, which brings us back to reality. Mixing is driven by the physical laws that dictate fluid motion, in the same way as the motion of solids is dictated by Newton's laws and, with respect to planetary motion, by Kepler's laws. But before we turn our attention to the physics involved, let us return to the two particular examples considered earlier with respect to survival in a stagnant atmosphere, breathing being the first and body-temperature control being the second.

When we breathe out in the real atmosphere, we exhale the air as a jet that propels the mass away from our face, see Figure 2.4. Described simply, this stale air is then replaced by fresh air that is being sucked in from the side towards our face — the blue arrows in Figure 2.4. Reality is somewhat more complex as regards the details, however. In particular, the jet we exhale mixes with the ambient air by the action of vortices that form within the jet and its edges. This process dilutes the jet to form a mixture of stale and fresh air. Some of this mixture may return to the vicinity of our face, so the air we breathe in will, in general, contain some of the mass we exhaled, in addition to a proportion of the stale air in the tail of the jet which contains the air exhaled last before we breathe in again.

The mechanism described here is called forced "convection", or more accurately "advection", a process in which fluid motion is generated by an

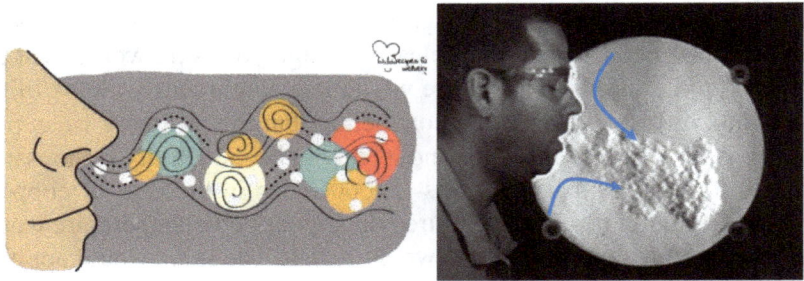

Figure 2.4.   Mixing between air exhaled from the lungs with ambient air by the action of turbulent eddies.

Source: Recipes for Wellbeing, www.recipesforwellbeing.org (with permission); National Institute of Standards and Technology, rights reserved (with permission).

external source — here, the pressure produced by our contracting lungs pushing air into the lower-pressure outside atmosphere.

As regards our ability to control our temperature, the essential mechanism at play here is buoyancy — the same one that is active close to the surface of central heating radiators in our home. When the air next to the skin is heated, it is being driven upwards because it is lighter than its lower-temperature environment. This is then replaced by cooler air that moves towards the skin from below and the side, as is shown in Figure 2.5. This is called "free convection" or "natural convection". Here too, the details of the process are more complicated.

First, as the air rises, a complex set of vortices is created within the rising layer, and this gives rise to mixing between the warm air and the cooler air outside the layer. This tends to lower the temperature in the layer, thus enhancing the ability of the body to liberate more heat. Second, the motion of the surrounding air and/or our body within it causes cool air to be transported by "forced advection" towards our skin. We are keenly aware of the importance of this process in hot weather through the welcome action of a cooling breeze or the air stream generated by a fan. The totality of this process is referred to as "mixed convection" — or more precisely, "turbulent mixed convection", the advective "turbulent" referring to the vortices within the warm layer. These vortices are also evident in the "forced advection" case shown in Figure 1.1. We discuss the subject of turbulence in much greater detail in Chapter 5.

As an interesting aside, the subject of breathing and, even more importantly, of coughing gained major prominence and public attention during the COVID-19 pandemic in the period 2019–2022. Critical

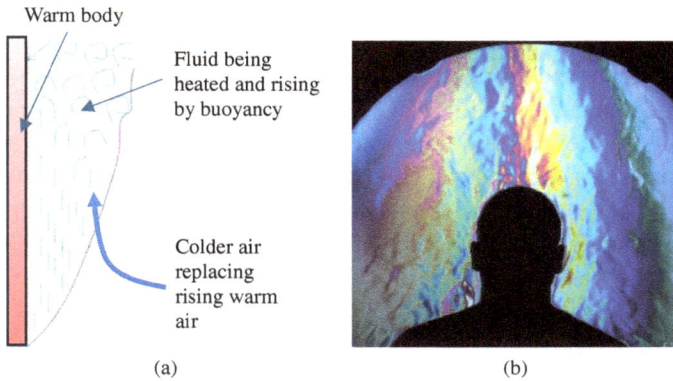

Figure 2.5.   Buoyancy-driven flow next to a heated wall or skin: (a) the fluid next to a heated wall or surface is rising upwards due to buoyancy; the layer next to the wall or body skin — called "boundary layer" — thickens due to mixing between the heated near-wall fluid and the cooler outer stagnant fluid which continuously replaces the heated air; (b) warm air rising from a heated body model.

Source: Settles [2] (with permission).

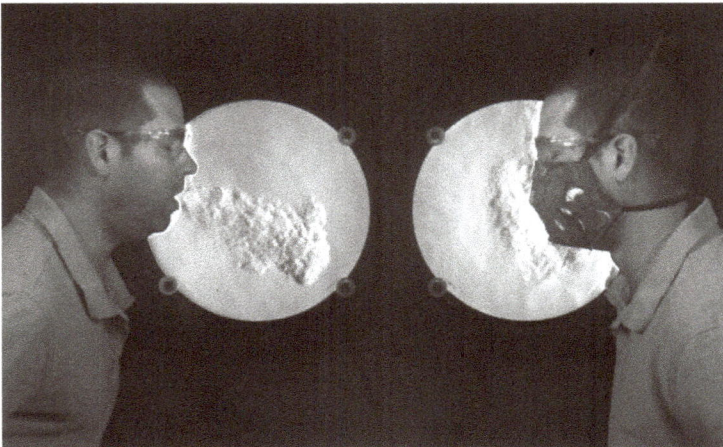

Figure 2.6.   Visualisation of coughing with and without a mask, the latter reducing the penetration of the jet of exhaled fluid.

Source: National Institute of Standards and Technology, rights reserved (with permission).

questions were asked, for example, about the extent to which masks prevented or hindered the spread of the virus, as is exemplified in Figure 2.6, and numerous technical investigations were undertaken on how masks diverted or diffused the stream of air being expelled by infected patients.

2.3 Life-Support Systems

Breathing serves the purpose of supplying our lungs, and hence our blood, with oxygen, and of expelling CO_2, a byproduct of our body's conversion of nutrients to thermal and mechanical energy. The air entering and leaving our lungs thus acts as a carrier of chemical substances, and its ability to deform freely within the air passages in our respiratory system is crucially important for the transport of O_2 and CO_2 towards and away from the blood-containing tissue in our lungs. The details of the physiological and physico-chemical processes by which this exchange of chemicals with the blood takes place are intricate and not relevant to this text. The essential feature to highlight here, by reference to Figure 2.7, is that the air passes through our nasal passage, itself quite complex, to or from the bronchi that form the membrane which separates the airways from the blood-containing tissue.

The bronchial system consists of several "generations" of bronchi dimensions, and the gas exchange with the blood takes place predominantly at the smallest size. This exchange depends crucially on the small-scale mixing within the air, which facilitates oxygen-rich air packages reaching the smallest bronchi.

The respiratory process shown in Figure 2.7 brings us neatly to the second existentially important fluid system in our body: the

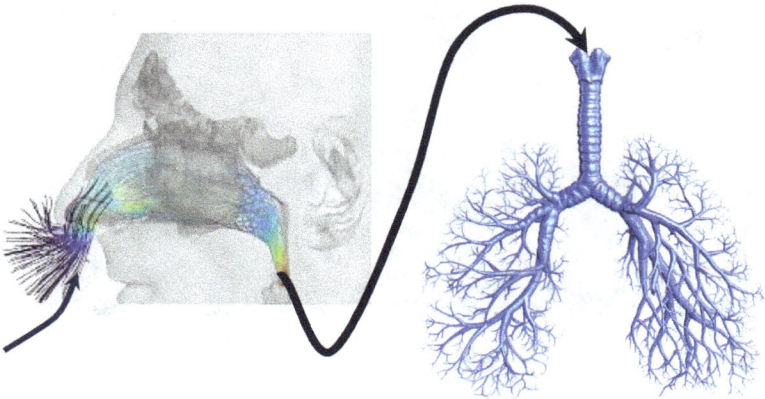

Figure 2.7.   The human respiratory process: air being sucked through the nasal passage into the bronchial system, which bifurcates through several "generations" (more than 20) of smaller and smaller size (the smallest being roughly 0.5 mm in diameter).

Source: M. Quadrio (personal communication); iStock (with permission).

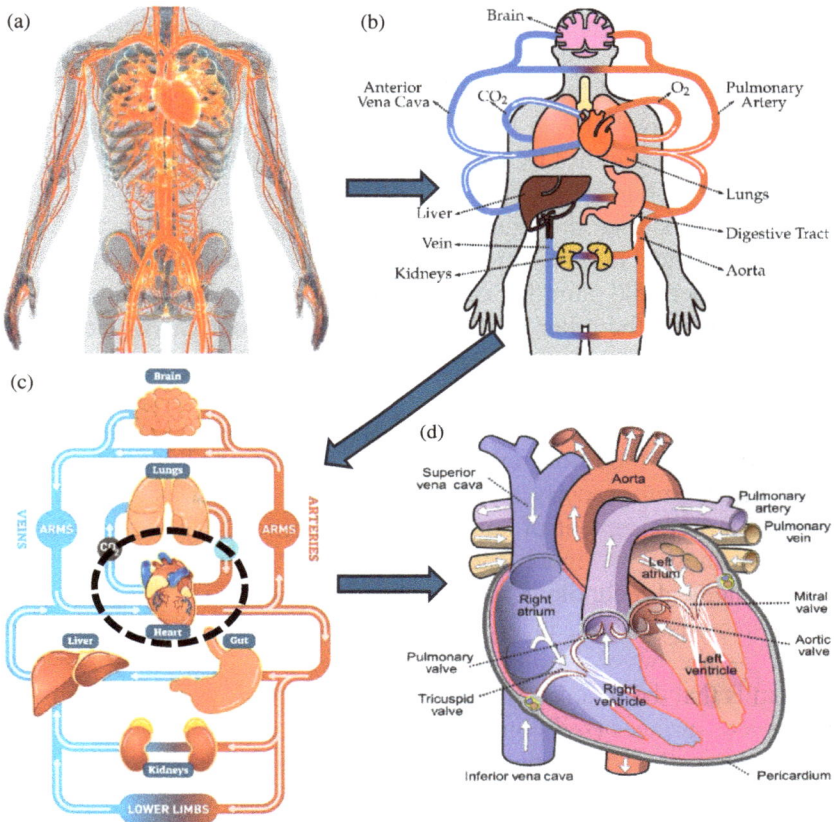

Figure 2.8. The cardio-vascular system — progressive abstraction: (b) and (c) progressive simplifications of the full system in (a); the heart and the flow within are highlighted in (d).

Source: iStock (with permission); Syed *et al.* [3] (with permission); Wikimedia Commons CC-BY.

cardio-vascular system, as shown in Figure 2.8. Here, too, the physiology is complex and its details are of marginal relevance. Indeed, even the system's geometric structure is far too convoluted to consider. What is most important to note here is that the heart periodically pumps oxygenated blood (red in Figure 2.8) from the lungs through the arteries into the essential organs of the body and returns the CO_2-enriched blood (blue) through the veins back to the heart and then to the lungs in which the air exchanges CO_2 for O_2.

The blood is a complex fluid containing numerous substances, among them red blood cells, platelets and white cells. However, in the present context, the essential feature to highlight is that the blood is a fluid that has to flow relatively freely through the intricate system of "ducts" — the blood vessels — and has to act as a carrier of nutrients, oxygen and CO_2 for life to be sustained.

The heart, shown in Figure 2.8(d), is an especially interesting fluid-mechanics device, discussed in greater detail in Section 6.11. Its function depends on periodic pumping of the blood in two circuits: one consisting of the veins and the other of arteries. Uni-directional valves ensure that the blood only flows in one direction, from the "atria" to the "ventricles" (the heart chambers).

Diseased valves — typically through the deposition of fatty tissue constricting the blood flow, or reverse-flow leaking resulting in the loss of the unidirectional property of the valves — are frequent causes of heart failure. The replacement of such valves by artificial valves, such as the examples shown in Figure 2.9, is one of the spectacular successes of modern medical technology and surgery. However, such valves — even the most modern ones, incorporating polymeric and pig-derived membranes — are never wholly satisfactory replacements. One particular problem is that the blood is subjected to strong shear as it flows across the mechanical surfaces (see Figure 2.9), and this causes physiological damage to the cells within the blood as well as undesirable turbulence

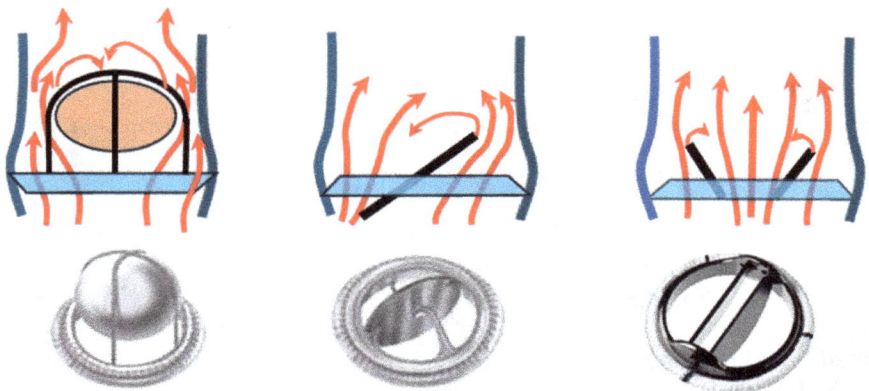

Figure 2.9. Examples of mechanical heart valves used to replace diseased natural valves and typical flow patterns around alternative moving gates.

(instability and eddies). We return to the fluid mechanics of alternative heart-valve designs in Section 6.11.

2.4 Whirling, Ascending and Mixing

There is an almost infinite variety of examples that can be fielded to illustrate the crucial role played by advection and convection — both being forms of transport — and mixing by vortices in flow configurations essential to our existence and well-being. We are surrounded by numerous flows, and we benefit from a wide spectrum of technological processes that involve these mechanisms. We need to develop a (modestly) deeper understanding of the laws governing fluid mechanics before embarking on a well-founded description of some interesting phenomena that require more than superficial observation; we do so in Chapters 4 and 5. However, as appetisers, consider chimneys, fires, volcanoes and our global weather.

The flow out of a chimney, shown in Figure 2.10, is a combination of a jet and a plume — i.e., a combination of forced and free convection.

Figure 2.10.   The discharge of warm steam from a chimney into a cooler atmosphere.
Source: © VistaCreate.

Figure 2.11.   A fire produced in a pool of liquid fuel. The yellow arrows indicate the entrainment of ambient air into the flame.

Source: Mishra *et al.* [4] (with permission).

It is driven upwards by the action of buoyancy. It accelerates as it is being channelled into the chimney stack, and it then issues into the atmosphere as a stream of fast, warm, moist fluid (or vapour). It is then deflected by wind — a process of forced advection in the lateral direction. Finally, the stream mixes with the surrounding atmospheric air by the action of turbulent vortices, causing the plume to spread and disperse. This last process is very similar to the breathing-out action shown in Figure 2.4.

A flame, such as that shown in Figure 2.11, is a more extreme version of the chimney plume. We ignore the details of the very complex combustion process, except to say that the flame is sustained by the mixing of oxygen-rich air with the fuel vapour. This occurs through the entrainment of air from the side and from below the flame, a process similar to the entrainment into the buoyancy-driven near-wall layer shown in Figure 2.5(a). Here again, the turbulent vortices play a crucial role in the mixing process. The flame then generates a hot smoke-containing plume,

(a)   (b)

Figure 2.12. Hot, dusty plumes created during volcanic eruptions: (a) plume with relatively little solid matter rising vertically due to buoyancy; (b) plume collapse and pyroclastic flow that contains a large amount of solids.

Source: A. Ionescu, Earth.com (with permission); pyroclastic flow, Encyclopaedia Britannica, 2024.

which rises into the surrounding atmosphere in the same way as the chimney plume in Figure 2.10 does.

The images of the volcanic eruptions shown in Figure 2.12 serve to illustrate a further set of phenomena that accompany forced plumes. Here, the hot plume is visualised by the presence of ash within the plume. The extreme temperature of the plume and the high upward speed that accompanies the eruption combine to drive the plume several miles upwards into the atmosphere, eventually spreading laterally over a wide area as it cools and stops rising further when buoyancy is lost.

The image in Figure 2.12(b) shows a much-feared phenomenon called "pyroclastic flow". This is a hot stream of dense ash, pumice and rocks that races down the slope of the volcano at speeds of up to 100 km per hour, the very phenomenon that destroyed Pompei following the eruption of Vesuvius (aided by the pyroclastic flow being able to race rapidly over the water of the Bay of Naples). This stream is formed when the upward buoyancy of the plume due to its high temperature is insufficiently strong to overcome the negative buoyancy due to the weight of the material within the edge of the plume. The edge of the plume then reverses direction and flows down at high speed, destroying practically everything in its path. A snow avalanche and a large landslide are both similar to a pyroclastic flow, although in these two cases, thermal effects play no role in the process.

2.5 Global Dance

With stellar and cosmological phenomena put aside, the global atmosphere probably represents the most elaborate amalgam of fluid-mechanic and thermodynamic phenomena we know. Although geologists concerned with subterranean fluid mechanics in the Earth's mantle and core might challenge this assertion, what is indisputable is that we know much more about what is above the Earth's surface than what is below it, thus being more aware of the former and able to appreciate the intricate interactions involved.

The weather is driven by a complex set of circulations, as shown in Figure 2.13, and an interplay between conditions on land and oceans, Earth-rotational effects, solar radiation, reflection, absorption, evaporation, precipitation and species transfer. In all these, the basic mechanisms of convection, advection, turbulence and buoyancy again play important roles. Discussing these interactions would be too ambitious at this early stage of the narrative. However, it is instructive to point to some phenomena with which we are familiar through daily observations and predictions reported in the media.

Wind is a manifestation of advection. At a scale of around 100–1000 km, it is driven by low- and high-pressure regions close to the surface, caused by ascending and descending air masses, as shown in Figure 2.14. A region of low pressure tends to draw into it air from surrounding higher-pressure areas because the rising air has to be replenished from the side. The reverse occurs in regions of high pressure: the descending air flow has to be forced outwards towards regions of lower pressure, for otherwise there would be an unlimited accumulation of mass at the surface. This basic connection between pressure differences and fluid motion (i.e., advection) applies generally: fluid is being driven from high to low pressure. A good example is air being driven out of a balloon by the high pressure inside the balloon. Another is a garden hose connected to an open tap.

Returning to Figure 2.14, you might be surprised to observe that the air tends to rotate as it flows towards the low-pressure area and away from the high-pressure area rather than move radially inwards or

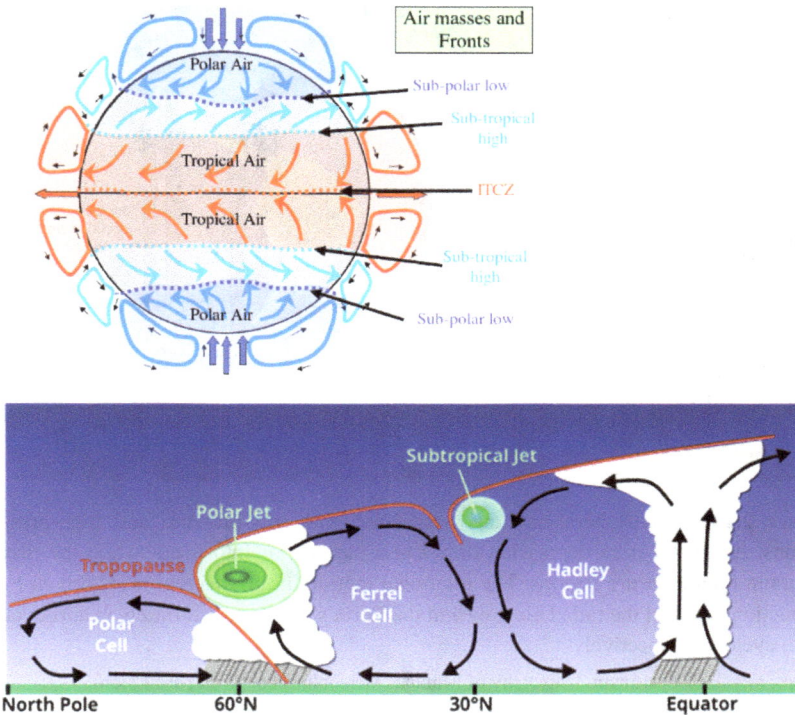

Figure 2.13. Major circulations and flow features around the globe.

Source: L. Urbano, Natural World, montessorimuddle.org (with permission); NOAA, National Weather Service, Wikimedia Commons CC-BY.

outwards. We discuss the origin of this rotation in Chapter 4 (Section 4.3) when dealing with mechanisms that give rise to the formation of vortices. Here, the only comment made is that this rotation is linked to the angular rotation of the Earth and a resulting force that deflects the air as it wants to flow radially inwards or outwards. This is also the reason why weather charts show the wind to move predominantly along "isobars" — lines of constant pressure — as is indicated in Figure 2.15.

Hurricanes and typhoons (though not tornadoes!) — generically referred to as "tropical cyclones" — are driven, essentially, by the same mechanisms as the process shown on the left-hand side of Figure 2.14. Such cyclones are formed over warm ocean surfaces, giving rise to strong updrafts and strong horizontal rotation over areas that extend to hundreds

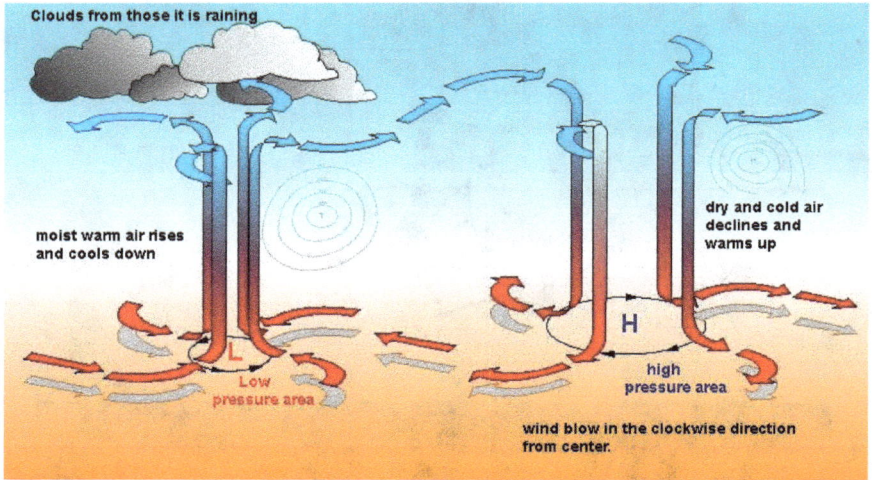

Figure 2.14. Atmospheric motions associated with low-pressure and high-pressure regions. A low-pressure depression draws near-surface air into itself, while the reverse occurs in high-pressure regions. The rotational motion in the two regions is driven by an interaction between the radial air flow and the Earth's rotation — forming an anti-cyclone and a cyclone, respectively.

Source: Physics Department, University of Munich; Meteoblue.com (with permission).

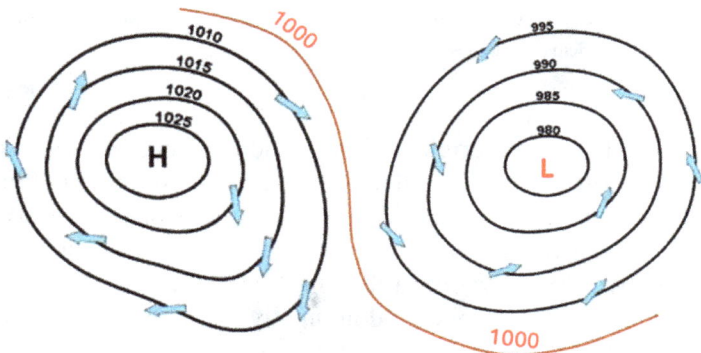

Figure 2.15. Typical wind motion around highs and lows in the northern hemisphere. Any line indicates a constant-pressure level or "isobar" (analogous to a constant-elevation line on a hill or in a valley). The wind is broadly aligned with the direction of the isobars because of the interactions shown in Figure 2.14.

Source: Meteoblue.com (with permission).

of kilometres, transporting huge amounts of moisture into the upper atmosphere and then resulting in extremely heavy rain. We discuss hurricanes and tornadoes in Chapter 4 when dealing with vortices.

At a much smaller scale, updrafts or "thermals" also occur due to local heating of the ground. Such thermals are exploited by gliders and birds to gain height. If sufficiently strong, thermals can manifest themselves by the formation of cumulus clouds, subject to the right combination of moisture and temperature conditions in the atmosphere. Any nervous flyer will be acutely aware of these updrafts and accompanying downdrafts as the plane passes through such cloud formations, especially during take-off and landing stages.

Yet another interesting situation, perhaps only observed by an inquisitive minority, arises in strong direct sunshine when packets of fairly hefty, rapidly moving clouds alternately obscure and expose the ground. The rapid heating and cooling of the ground lead to local updrafts and downdrafts that provoke horizontal motions that can be quite strong. In contrast to the large-scale regions shown in Figure 2.14, these lateral motions are not accompanied by more than marginal rotation because the forces arising from the Earth's rotation are too weak to be effective at spatial scales of a few tens or even hundreds of meters. They are also mostly too short-lived to develop into structures akin to tornadoes and dust devils, the mechanism of which is discussed in Chapter 4.

The fluid mechanics of large seas and the oceans are no simpler than those of the atmosphere but are driven by a number of mechanisms that are different to those in the atmosphere. One obvious difference arises from the constraining influence of the continents. These define the extent of large-scale circulation patterns, as shown in Figure 2.16. These patterns feature warm currents (red arrows) that flow from tropical zones towards temperate and subpolar zones, as well as cold currents (blue arrows) from polar regions towards the equatorial region. It needs to be pointed out that the pattern shown in Figure 2.16 is a highly simplified (but readily digestible) representation of reality which features a more complex set of circulations and cells that are present in the oceans.

Other important mechanisms that are different from those driving the atmosphere include wind-induced shear on the water surface that generates both horizontal and vertical motions and causes waves and turbulent eddies, complex ocean-floor topography, and buoyancy effects due to density differences that arise from solar radiation and temperature and salinity variations caused by evaporation, cross-surface heat exchange, sub-surface mixing, terrestrial-water runoff and ice melts in the polar

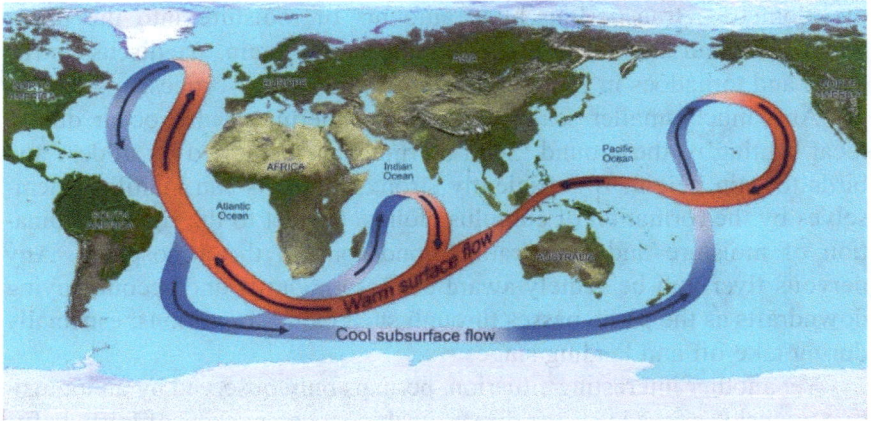

Figure 2.16.   Ocean currents across the globe around continents. Red arrows identify warm ocean currents that originate near the equator and move towards the poles or higher latitudes, while blue arrows identify cold currents that originate near the poles or higher latitudes and move towards the tropics or lower latitudes.

Source: NASA/JPL.

regions. The Earth's rotation is also influential, but this is one driver that large aquatic bodies share with the atmosphere, the mechanism being the same as that described earlier by reference to Figures 2.14 and 2.15.

Wind shear has an especially powerful influence on oceanic fluid mechanics. It generates large waves, strong surface-proximate currents, depth-wise currents and turbulence, all of which promote the mixing of properties of the water in the vertical direction; we consider the subject of waves in Chapter 6. Wind-induced currents and turbulence in the ocean are also crucially important to aquatic life. In particular, the interaction between the wind and the sea surface has a massive influence on the exchange of oxygen and CO_2 across the surface, with turbulence redistributing the properties of the water in the vertical direction through the action of mixing.

2.6 Closing Remarks

This chapter has endeavoured to give a first flavour of the importance of fluid flows to our existence and the world in which we live. This was done, principally, by way of specific flow scenarios from the natural world most of which will be reasonably familiar to the majority of us.

While all flow examples included in this chapter have been described in fairly superficial terms, a key message that is important to take away, from a fluid-mechanic perspective, is that advection (mass motion forced by "pushing", "pulling", pressure differences and geometric obstructions), convection (mass motion driven by buoyancy), mixing and vortices are all major constituents of almost all fluid flows. This applies to the natural environment as well as to most engineering flows, the latter yet to be discussed in chapters to follow.

Most natural and environmental flows are forced upon us by the way the natural world is configured. We may be able to exercise some degree of control over their properties and characteristics by means of creative engineering solutions (e.g., dams, canals, levees, flood-plain zoning, wave breakers and windshields), but on the whole, we have to accept them and learn how to live with them. In contrast, fluid-flow engineering harnesses the properties of fluids to achieve a desirable performance. To do so effectively, we have to understand what makes fluid flows "tick" — i.e., we have to understand the interactions between agencies and forces that drive flows and the way in which the flow properties respond to external forcing, geometric features, control measures, the transfer of heat and mass in and out of flows and the response of flows to the addition of particles, foreign species and bubbles.

What we do next is to go through a brief introduction to techniques that allow the dynamic characteristics of flows to be measured and simulated, followed by a largely descriptive overview of the major rules that dictate the physical behaviour of flows.

Chapter 3

Unmasking the Hidden Faces
and Interior of Flows

Throughout this book, we encounter many pictures and images that indicate what happens within fluid flows and are used to support explanations of how flows behave and how they interact with their surroundings or with solid boundaries confining them. How are these images obtained and how do we elucidate the properties of the flows shown in the images? The purpose of the current chapter is to answer this question, as a precursor to our exploring a variety of flow scenarios in the chapters to follow.

The large majority of fluid flows — and hence their internal structure and mechanics — are hidden from us. In fact, this is one important reason why so many among us have only a poor appreciation of the processes inside fluid flows. In a few cases, we can readily observe certain features — for example, the propagation of waves on liquid–air interfaces. In some other flows, the presence of smoke, vapour, glowing gases, plasma, bubbles or particles makes the flow visible — e.g., in a chimney discharging steam (Figure 2.10), in flames (Figures 2.1 and 2.11), in a volcanic eruption (Figure 2.12) or in sprays discharged from a can. In some atmospheric and aquatic flows, we may be able to infer certain characteristics by observing the effects of fluid motions on flexible structures — e.g., trees, bushes and flags in windy conditions, pieces of paper and leaves being carried by wind gusts or the movements in the canopy of a kelp forest in ocean currents. However, in almost all of these cases, the information we gain is purely qualitative and highly fragmentary in nature. We can certainly see very little that conveys an understanding of the processes within the flows, and we can derive little or no quantitative information from such observations.

If we wish to "see" into a flow, we have two main options: perform targeted experiments or use computer simulations to predict the flows. These are not really alternative techniques but are, rather, complementary in terms of the type of output they provide. Thus, we often have to employ both for one and the same flow to gain a comprehensive picture of the flow-physical processes at play or to make sure the two methods give consistent statements, especially when we have doubts about the veracity or accuracy of a simulation.

3.1 Observing and Measuring

At a very basic experimental level, we can use smoke visualisations (see Figure 4.3 in Section 4.1). In these, the smoke particles are carried by the flow and the traces — called "streaklines" or "streamlines" — indicate the path taken by fluid packets. Importantly, these traces also indicate the direction of the velocity inside the flow, this being tangential to the traces (the velocity is defined as the speed — the magnitude — and the direction of the fluid motion). As is explained in the following chapter, the streak-lines allow us to figure out, in qualitative terms, where the flow is accelerating and decelerating and where the pressure is rising or falling. However, this approach is subject to serious limitations. If the flow is turbulent, the smoke trails meander wildly and quickly disappear due to mixing and dilution. Also, we can only focus on two-dimensional planes, and it is almost impossible to visualise, beyond a very superficial level, flows that are strongly three-dimensional in character, for example, when we aim to investigate the effects of buildings on wind conditions in urban planning. "Schlieren" images, of the type shown in Figures 2.2 and 2.4–2.6, rely on the refraction properties of light as it passes through regions of variable fluid density. Again, this method provides little more than a qualitative impression of the flow.

There are several experimental techniques that allow us to measure the velocity (speed + direction) in flows. One is called "hot-wire anemometry". This involves the insertion of a probe, on which a very thin wire is mounted, into the flow, as shown in Figure 3.1(a). The wire is heated electrically. The cooling of the wire by the flow at the location of measurement leads to a change in electrical resistance in the wire, which can

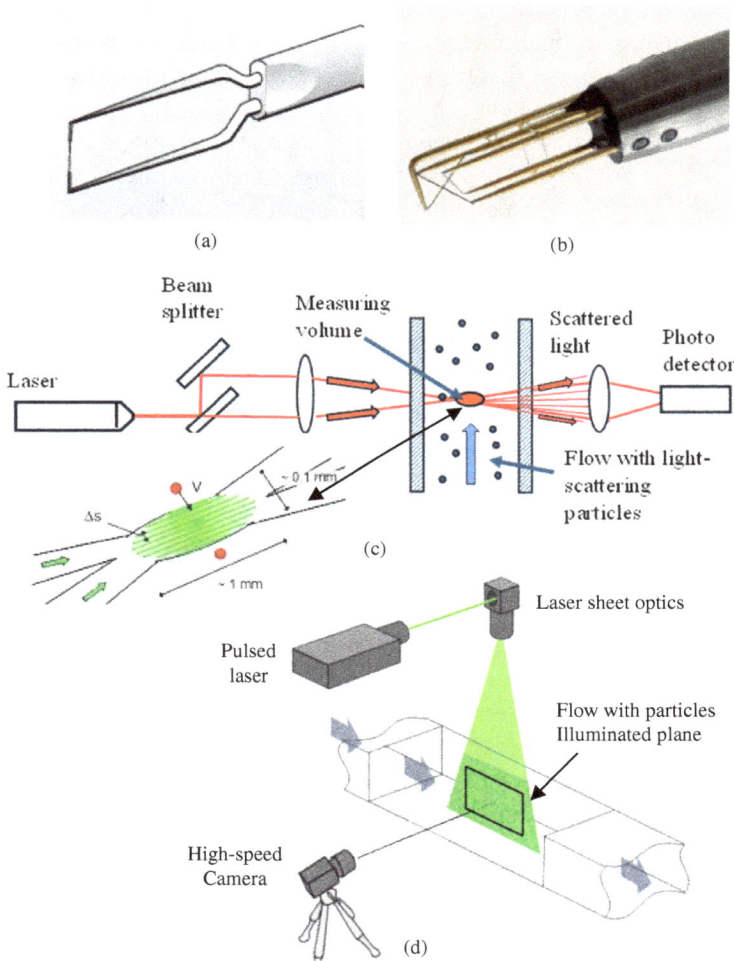

Figure 3.1. Three experimental approaches to measuring flow velocity: (a) single hot-wire probe, used to measure the normal-to-wire velocity; (b) multi-sensor hot-wire probe, used to measure three velocity components; (c) laser-doppler anemometry; (d) particle image velocimetry.

Source: Correia *et al.* [5] (with permission); © Dantech Dynamics (with permission).

be translated into the speed normal to the wire at its location. More elaborate sensors combine several hot wires mounted in different orientations, as shown in Figure 3.1(b), thus enabling different velocity components to be measured simultaneously.

A second technique is called "laser-doppler anemometry". This is a non-intrusive optical method, the major advantage being that the optical sensor does not disturb or interfere with the flow. As is shown in Figure 3.1(c), two laser beams are made to cross at an angle at the location of measurement. An interference pattern is created within a small volume at the intersection of the beams. The fringes are separated by a distance that depends on the wavelength of the light source and the angle of beam intersection, both known. As small particles (e.g., dust or tiny oil droplets injected into the flow) are carried by the flow, they cross the fringe pattern, and an optical detector counts the number of fringes crossed by the light-reflecting particles in time, giving information on the speed normal to the fringes.

A third technique is called "particle-image velocimetry", as shown in Figure 3.1(d). Here, optical images of particles within a portion of the flow field are taken at two instances in time separated by a short interval. The two images are then correlated and used to infer the displacement of the particles in the time interval, and this allows the velocity field in the optically resolved area to be determined.

All three methods are subject to significant limitations in terms of the flexibility of varying the geometry and the dynamic conditions of the flow, the level of resolution, and the cost of the instrumentation, the wind tunnels and the test geometry containing the flow. The third method is the most elaborate and the least restrictive, in so far as it can be used to map reasonably large portions of three-dimensional flow by using a stereoscopic arrangement of cameras. However, this technique is also extremely costly.

One further, very important, limitation of the techniques discussed so far is that none is able to measure the pressure inside the flow. The pressure is important, however, not only inside the flow but also at solid boundaries, because it gives rise to forces — for example, those responsible for lift and drag on aircraft wings and cars. As is explained in Chapter 4, spatial differences in the pressure within the flow are also of major interest because they are intimately linked to fluid acceleration, deceleration, turning and twisting.

At solid boundaries, pressure can be measured, albeit only at chosen point locations, by means of small holes drilled into the surface, which are then connected with thin tubes to a mechanical or electronic manometer. More advanced alternatives are pressure transducers, which convert pressure into electrical signals, or pressure-sensitive paint, the colour of which changes with the pressure. However, the latter method requires elaborate calibration, optical instrumentation and data-processing techniques. In addition, its accuracy is sensitive to temperature. In contrast, the discrete-hole method is simple and fairly accurate but relies on the flow being tangential to the surface. If, on the other hand, the flow is an angle to the surface, the measured quantity is not the true pressure — conventionally referred to as the "static pressure" — but, rather, the sum of this pressure and an increment that arises from the impingement of the flow onto the surface, due to the normal-to-surface flow being decelerated to zero at the surface. This is most easily visualised by placing a plate (or the palm of your hand) against a jet. The jet impinges on the wall, decelerating and being pushed sidewards, and this results in an additive "dynamic" increment that is not part of the real "static pressure". Chapter 4 provides further explanations to clarify the relationship between the pressure and the flow field.

The pressure inside the flow can be measured with physical tube-based probes inserted into the flow. The most widely used probe is called the "Pitot tube", its principle being conveyed in Figure 3.2(a). To avoid undue misrepresentation, it needs to be pointed out that the main purpose of a pitot tube is to provide information on the flow speed, the static pressure being one of the quantities necessary to derive the speed. In fact, Pitot tubes are used on most aircraft, as exemplified in Figure 3.2(b), to measure the speed of the aircraft relative to the ambient atmospheric conditions. When carefully aligned with the flow direction, a Pitot measures two quantities: the pressure on the tube wall, via a hole in the wall, and the pressure in an opening facing the flow. The former is the static

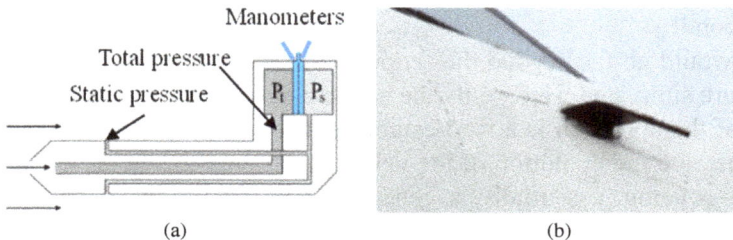

Figure 3.2. Pressure- and velocity-measuring Pitot tubes: (a) operational principles; (b) use on an aircraft body.

Source: Ezzeddine *et al.* [6] (with permission).

pressure, while the latter is the sum of the static pressure and the impingement-related "dynamic" pressure already mentioned above. The difference between the two thus provides information on the speed along the tube, via a relationship explained in Chapter 4. Here, we merely need to note, without proof, that the dynamic pressure is proportional to the square of the speed.

Apart from the fact that Pitot tubes have to be carefully aligned with the flow direction — easy to do on an aircraft but much more difficult in a complex three-dimensional internal flow — Pitot tubes distort the flow, by their physical bulk, thus causing measurement errors. This is a disadvantage of all intrusive measurement techniques, avoided with the two optical velocity-measuring techniques shown in Figure 3.1(c,d).

3.2 Modelling and Number-Crunching

An entirely different and extremely powerful approach, now used extensively in physics and engineering, is one that takes advantage of computer codes to model or simulate flows — yielding a virtual reality of the flow. This approach is referred to as "Computational Fluid Mechanics". Most of us see one particular outcome of this method daily on our TV screens in the form of weather forecasts projected over a period of several days. This branch of fluid mechanics has benefitted greatly from a phenomenal increase in computer speeds over the past two to three decades. It is hard to comprehend the number-crunching speeds sustained by the most powerful supercomputers in use. At the time of writing (2024), the most powerful machine — the "El Capitan", located in the Lawrence Livermore National Laboratory in California — is capable of performing over 10^{18} ($10^6 \times 10^6 \times 10^6$ or million × million × million) floating-point operations per second!

It would go far beyond the scope of this text to explain in detail how flows are simulated. Here, only the barest essentials can be described. The basis of the approach is a well-established set of mathematical equations that describe the evolution of the velocity and pressure fields in time and space — being, essentially, a general expression of the widely known relationship between force and acceleration — i.e., Newton's laws of motion; we return to this topic in Chapter 4 to convey a better appreciation of what is involved. These equations are quite general and apply across many flows in nature and engineering. Unfortunately, they are too complex to be solved analytically (with "pen and paper") and require the use

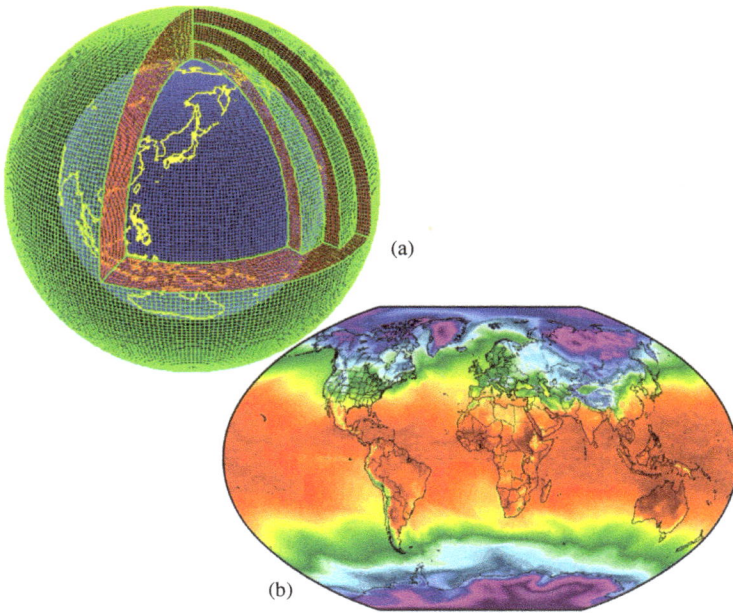

Figure 3.3. Numerical prediction of the global weather: (a) numerical mesh; (b) surface temperature on Sept. 4th 2010.

of approximate numerical techniques. In essence, the equations are reformulated to give discrete values of velocity and pressure at the "nodes" of a mesh spanning the flow. As is shown in Figure 3.3, the mesh used in weather forecasting may cover the whole globe, from the ground to a height of several miles, or a smaller geographic area — a continent or a country. In the case of engineering flows, such as the car and aircraft bodies shown in Figures 3.4 and 3.5, respectively, the mesh extends from the solid surface of the body to the outer region in which the flow conditions are undisturbed by the body and can be assumed to be known and can thus be prescribed as "boundary conditions". The conditions at each mesh node — velocity, pressure, temperature and water content, among others — are coupled to those at neighbouring nodes. All the coupled equations at all the nodes (which can number several billions) are then solved together on the type of supercomputers the speed of which was indicated in the previous paragraph. The set of velocities at the discrete nodes can then be

Figure 3.4. The numerical mesh around an F1 car, and the outcome of a computer simulation in the form of streaklines and surface pressure indicated by colour.

Source: cfmesh.com (with permission); ANSYS Ltd. and Verus Engineering (with permission).

used to construct flow pictures of the type shown in Figures 3.4 and 3.5. Indeed, if we solve the equations repeatedly at consecutive time levels, we can produce movies that show how the flow in question evolves in time. Such results would be virtually impossible to derive experimentally, let alone at an acceptable cost.

Alongside the velocity field, the pressure field is also resolved. As explained earlier, the pressure is very difficult to obtain experimentally in any detail and with sufficient accuracy, except at solid boundaries. In fact, the temperature field and the distribution of any chemical species can also be determined if additional equations expressing the transport of heat and material species are incorporated into the computer code, as is done in weather prediction and the modelling of flames, for example. Chemical-reaction models used for predicting flames may involve dozens of reactions and related chemical species, and this necessitates the solution of dozens of equations, in addition to those for the velocity, pressure and temperature.

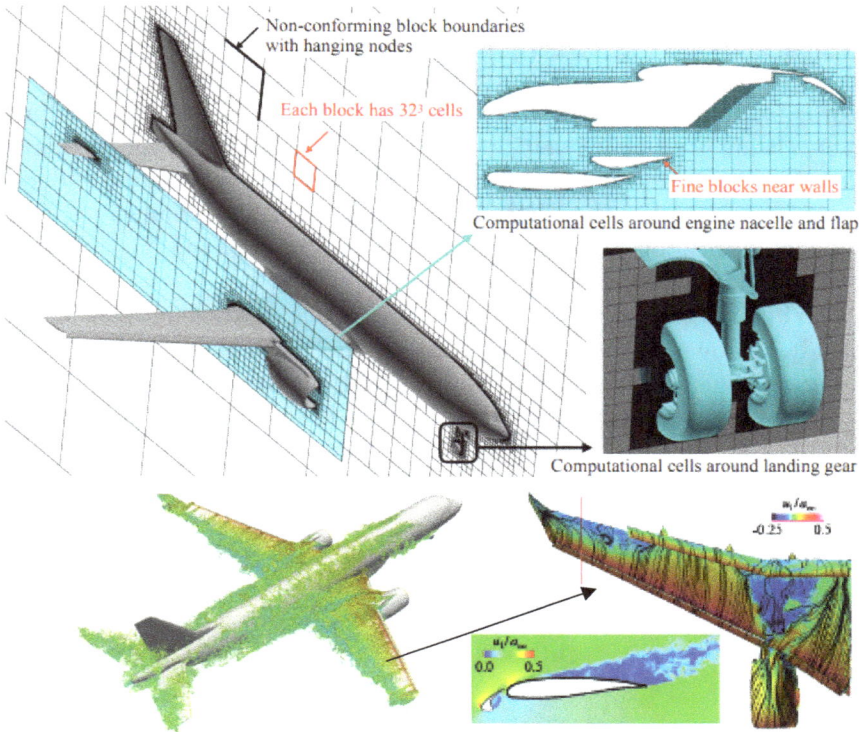

Figure 3.5. The numerical mesh around the body of a civilian airliner and the predicted flow close to the aircraft body, wing and engine surface in near-stall conditions with slat and flap extended.

Source: Asada *et al.* [7], © 2023 by authors (with permission).

There are literally hundreds of laboratories and research groups in universities, research establishments and industry engaged, worldwide, in experimental and computational studies of fluid flows. Many engineering components, machines, engines and whole systems are now designed primarily on the basis of computer predictions, which are usually verified or validated by means of limited wind-tunnel testing. This is now the standard approach used in many industrial sectors, including those manufacturing aircraft, jet engines, cars, high-speed trains, ships, wind turbines and energy conversion equipment. As noted already, weather forecasting is now based on continuous simulations performed by national weather centres on some of the most powerful multi-million-dollar computers in existence. As the simulations evolve in time, measurements at key geographic locations are regularly fed — assimilated — into the predictions to correct and guide the computations as the weather conditions are extrapolated over longer and longer time intervals. Computer modelling is also being used extensively to predict marine and river flows, including flooding scenarios, building aerodynamics, including ventilation and air-conditioning, underground water flows, metal casting, liquid-food processing, lubrication and many other industrial, natural and physiological processes that involve fluids.

While we cannot deal with the mathematical framework that underpins fluid-flow simulation codes, what we shall do in the following chapter is to discuss, in largely descriptive terms, the basic physical concepts and laws that ultimately describe the kinematic and dynamic properties of fluid flows.

Chapter 4

Minimalist's Fluid Mechanics

Readers of this book will, most likely, be familiar with Newton's laws of motion, applicable to a solid mass. Two are recalled below, because they are also relevant to the motion of fluids:

- A mass moving in a given direction will maintain its speed and direction unless a force is applied to it.
- A force applied to a mass results in the mass being accelerated in the direction of the force, the acceleration being proportional to the force and indirectly proportional to the mass.

Some qualifications and extensions may be helpful for discussions to follow:

- A mass moving with a particular speed is said to have "momentum", which is the product of the mass and the speed.
- A change in momentum of a mass involves a change in speed, and this change is proportional to the force and the period of time over which the force is maintained.
- A mass rotating with a particular speed at a given radius from the centre of rotation is said to have "angular momentum", which is the product of the mass, the speed of rotation and the radius. Thus, if the rotating mass is forced to move radially inwards or outwards, its speed of rotation has to increase or decrease, respectively, in order for the angular momentum to be conserved.

- A change in the angular momentum of the mass requires the application of a torque. This is analogous to a force causing a change in the linear momentum of a mass.

One further qualification that might appear to be superfluous, at this stage, is that the solid body to which the above statements apply is assumed to have a constant mass.

Newton's laws of motion also apply, in principle, to fluid bodies. However, there are some subtle differences that will become clear in the following sections. One difference is that the quantity equivalent of the force in a fluid flow is the pressure. In fact, a force is simply the pressure multiplied by the area on which the pressure acts. Another difference is that, in a fluid flow, we do not normally focus on a fluid body of a given mass but, rather, on a mass flow per unit time — the mass flux — more precisely, the amount of mass flowing per unit time through a particular area. It is this particular difference between a solid body and a fluid stream that is the focus of the first section to follow.

Before we embark on the following simple fluid-mechanic considerations, it is important to point out that the purpose of these considerations is to support the description and interpretation of the phenomena observed in flow examples to follow. In particular, the behaviour of a flow is dictated by the effects of pressure on the velocity and momentum of the mass within the flow. Hence, unless the fundamentals of this interaction are explained, the features observed in flow images cannot be explained clearly and thus appreciated.

4.1 Mass Cannot Disappear

This subtitle should be self-explanatory to all — unless one is in the realm of nuclear physics where mass is converted to energy. Still, it requires further elaboration in the context of fluid flows in order to appreciate the implications of observations to follow.

In solid mechanics, the concept of "mass conservation" is simple: a lump of material subjected to all sorts of forcing, deformation, movement, etc. is conserved. Solid fuel might appear to disappear, but it does not: it reacts with oxygen and is converted into gas and possibly soot during the combustion process. Ice melts to become liquid or water vapour (by sublimation). Here, it is relevant to highlight that we have already entered the realm of fluids — gas, clouds of soot, water and vapour.

In principle, the same rule applies to fluids, but we have to be more judicious in describing mass conservation with sufficient rigour. The problem with focusing on an identifiable lump of fluid is that mass is flowing continuously through any geometric configuration of interest. Mass is flowing in, flowing through, flowing out again and disappearing into the far distance. Hence, what we need to do is to consider mass conservation in terms of the *amount of mass per unit time* that flows in and out of a given volume. We call this "mass-flow rate" or "mass flux" (e.g., in kilograms per second, in metric units).

Consider the case of a room model which has three windows, as shown in Figure 4.1. By the action of drafts, air enters one window and streams out through two others. As an aside, it is instructive to note that this requires a pressure difference between the front and the back, which drives the air through the room. More about pressure later, however.

The mass-conservation principle can now be described as follows:

$$(\textit{mass flux in}) = (\textit{mass flux out})_1 + (\textit{mass flux out})_2$$

Since

$$\textit{mass} = \textit{volume} \times \textit{density}$$

it follows:

$$\textit{mass flux} = \textit{volume flux} \times \textit{density}$$

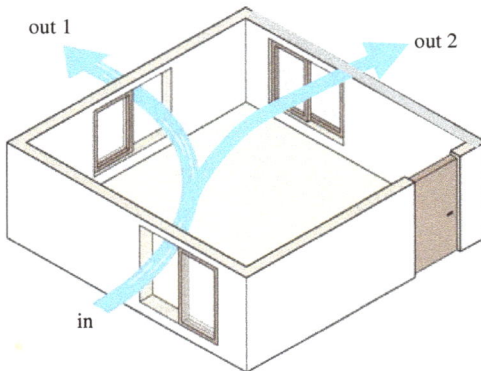

Figure 4.1.   A room model with a stream of air entering one window and its components exiting through two other windows.

Source: iStock (with permission).

where "volume flux" is the volume of fluid per unit time.

Also, since

volume flux = (*area of flow*) × (*distance the fluid travels per unit time*)

and *distance per unit time* is the speed (or *velocity*), it follows that the mass-conservation principle applicable to our room model can be expressed as follows:

$$\begin{aligned} density \times (flow\ area)_{in} &\times velocity_{in} \\ &= density \times (flow\ area)_1 \times velocity_1 \\ &+ density \times (flow\ area)_2 \times velocity_2 \end{aligned}$$

Thus, in general,

mass flux = *fluid density* × *flow area* × *flow velocity*

To be precise, we need to qualify here that we assume that the velocity does not vary across the flow area and that the velocity is at right angles (normal) to the area. If the velocity is not uniform, we need to interpret the velocity as being the average value across the area.

Let us consider two simple examples by means of which we can illustrate the implications of the above statements. The first is the narrowing tube, as shown in Figure 4.2. There is only one inlet and one outlet. The walls are impervious — i.e., the mass flux along the tube is constant. Assuming that the density of the fluid is also constant (say, water), we conclude

$$(area \times velocity)_1 = (area \times velocity)_2$$

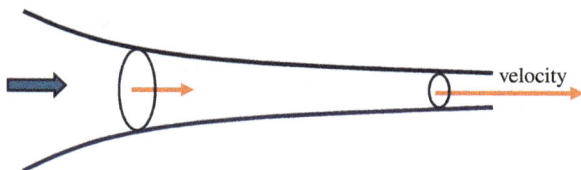

velocity

Figure 4.2.   The flow in a narrowing tube.

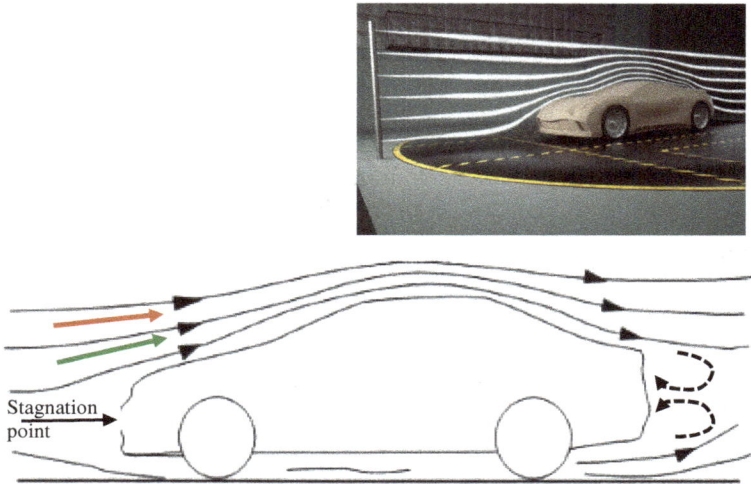

Figure 4.3. The flow around a car body, visualised by smoke trails and equivalent streamlines.

Source: iStock (with permission); Sivaraj *et al.* [8] (with permission).

which means that the velocity must increase in proportion to the reducing tube area.

The second example is the car-body model shown in Figure 4.3. In fact, as will transpire, this example represents an extension and generalisation of the previous example. The sketch, assumed to show the flow on the symmetry plane of the car, is a simplified view of a wind-tunnel experiment (the photographic insert) in which smoke trails are used to visualise the aerodynamic characteristics of the flow around a real car body.

In making the transition between the two examples, we need to first negotiate a modest logical hurdle. This hurdle is one of accepting that any streamline in Figure 4.3 is (almost) equivalent to a solid wall of the type in Figure 4.2. A "streamline" — a more precise term is a "streakline" — is a curve to which (by definition) the velocity is tangential — e.g., the red and green arrows in Figure 4.3. This should not be too hard to accept if we imagine that we ride on a smoke trail. As the flow turns, we turn with it, and to do so, we have to reorient our direction of travel to follow the trail. It should thus be obvious that no mass crosses a streamline, for otherwise, such a mass flow would necessitate a velocity component normal to the streamline and thus normal to the velocity directions shown in Figure 4.3. But this would fundamentally invalidate the concept of the streamline and the tangency of the velocity vector to it. We conclude, therefore, that any pair of neighbouring streamlines share with any equivalent pair of solid walls a common property: the pair bound a stream of constant mass flux. It then follows that the velocity between two converging streamlines must increase in proportion to the convergence, while the velocity decreases when the streamlines diverge.

Returning to Figure 4.3, we can now make some interesting observations regarding the flow around our car body:

- As the flow approaches the car (towards the "stagnation point"), the streamlines diverge, and the air velocity progressively decreases (down to a nominally zero velocity at the stagnation point).
- Upstream of the wind shield, the streamlines also diverge, though less severely, so the air velocity also decreases, but the velocity increases further downstream towards the roof — i.e., the flow is accelerating.
- Over the roof, the streamlines are highly bunched, so there, the air velocity must be high. As the flow progresses to the rear, the streamlines diverge again, and the flow decelerates.
- Behind the car, the streamlines are ill-defined, and we can conclude little about the flow field purely from the highly diffuse smoke trails and streamlines. What we do know, however, from many other observations and computational predictions, is that the flow in this region is, on average, slowly recirculating "eddying". This region is occasionally (and wrongly) referred to as a "dead-water/air zone". The recirculating motion behind the car is responsible, among others, for road dirt being lifted and transported towards the rear of the car and being deposited on the rear wind screen and the boot cover.

The above considerations, although made specifically in the context of car aerodynamics, are quite generic and widely applicable. To exemplify this point, attention is directed to Figure 4.4 which shows the flow over an idealised, long hill in a steady wind. As the flow approaches the hill, it is diverted upwards and accelerates as a whole because the flow area between the ground and the outer undisturbed (or less disturbed) flow narrows. Near the windward corner, the local flow decelerates, however, because the hill represents an obstruction to the oncoming stream. This is analogous to the flow just above the stagnation point in the front of the car model in Figure 4.3.

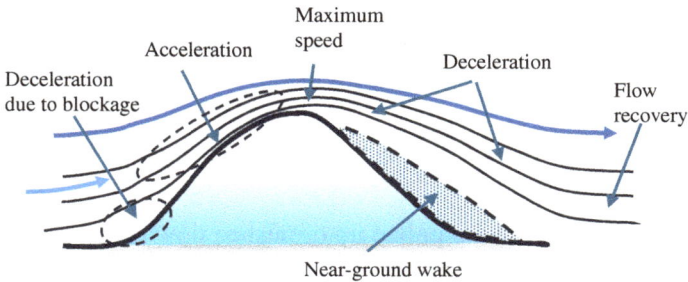

Figure 4.4. The flow over an idealised hill in constant wind. The hill is assumed to be long (deep), normal to the page (i.e., a ridge), so that the flow conditions in the cross-section of the hill can be considered representative of those at any other location along the hill.

As is explained in the following section, the pressure at this position is high, due to the blockage at the corner. This pressure drives the flow up the hill, and the acceleration in this region is characterised by the bunching of the streamlines. The bunching is highest at the top of the hill, as it is above the car body in Figure 4.3, and the wind speed is therefore highest at this location. Climbers will be well aware of the fact that the wind speed is fiercest at the crest of a hill or mountain in stormy conditions. The flow behind the hill decelerates as the flow area broadens, and the streamlines diverge, as they do behind the roof of the car body in Figure 4.3.

Next to the leeward side of the hill, there is a shaded region identified as the "Near-ground wake". The flow conditions in this wake require us to discuss mechanisms that are covered in Sections 4.4 and 4.5, and we leave this flow portion aside for later consideration. Suffice it to say here that the flow in this portion shares some features with the flow in the recirculating wake behind the car body.

4.2 Pressure Drives Speed (and Vice Versa)

The mass-flow considerations in the previous section allowed us to understand how we can infer changes in velocity in the flow by examining the convergence and divergence of the streamlines. However, we can go much further if we are prepared to examine, even in highly simplified terms, the relationship between pressure-induced forces and the decelerating or accelerating flow. It turns out that these considerations allow us to infer how the pressure varies in the flow, why our car needs to be driven forward to counteract the push-back — i.e., drag — by the flow, and why particular

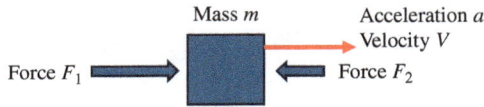

Figure 4.5.   A mass accelerating by a net force action towards the right.

portions of the car are being pulled up or pushed down, possibly requiring a spoiler to be installed at the back of the car to counteract the upward pull.

As we did in the previous section on mass conservation, we start our consideration of the relationship between pressure, acceleration and mass flux with (hopefully) familiar solid-mechanics concepts. For the sake of simplicity and brevity, we are going to use symbols that will be easily recognised as being obvious representatives of associated words.

In Figure 4.5, we consider a mass m subjected to a force F_1 acting on the mass from the left and a smaller force F_2 acting on it from the right. Newton's second law of motion tells us that the mass will accelerate at a rate a towards the right with an increasing velocity V. The relationship is

$$F_1 - F_2 = m \times a$$

If the net force $(F_1 - F_2)$ is applied suddenly at time t_1, when the body is already moving at velocity V_1, and the net force is maintained until a later time t_2, then the acceleration is constant and the velocity increases to V_2. The relationship is then

$$F_1 - F_2 = m \times (V_2 - V_1)/(t_2 - t_1)$$

i.e., the acceleration is equivalent to the rate of change (in time) in the velocity. Conventionally, we call *mass × velocity* "*momentum*", denoted in the following by M. Hence, the net force is equal to the rate of change of momentum:

$$F_1 - F_2 = (M_2 - M_1)/(t_2 - t_1)$$

How do we now translate all this to fluid flow?

We consider again the tube flow previously shown in Figure 4.2 and focus on a volume of fluid identified by the broken lines in Figure 4.6. The mass-flow rate — the mass "flux" — in the tube is obviously constant. We denote this steady flux through the volume by m_t. Here, it is important to reiterate that the mass flux is mass flow per unit time, e.g., kg per second in metric units. Dimensionally, this is equivalent to

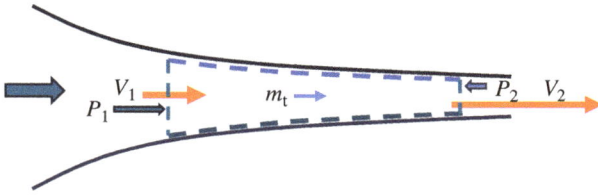

Figure 4.6. The flow in a narrowing circular tube; the broken line defines a portion in which the flow accelerates and on which pressure forces act.

$m/(t_2 - t_1)$ in the above solid-mechanics equivalent (Figure 4.5). The pressures P_1 and P_2 acting on the left and right faces are equivalent to the forces F_1 and F_2, respectively. In fact, the forces acting on the faces are simply the pressures multiplied by the respective areas.

Purely for the sake of transparency, we are now going to make an incorrect simplification — namely, that the duct walls do not apply any force on the fluid in the axial direction. This is incorrect because roughly the average between the pressures P_1 and P_2 along the duct wall acts on the area projection $(A_1 - A_2)$ of the fluid body against the flow direction. Subject to this simplification, the equivalent relationship to the force–acceleration argument in solid mechanics is now

$$net\ force = P_1 \times A_1 - P_2 \times A_2 = (mass\ flux) \times (V_2 - V_1)$$
$$= m_t \times (V_2 - V_1)$$

A simple example of this force–momentum relationship is the jet engine shown in Figure 4.7. Regardless of the complexities of the engine itself, a global statement that can be made is that the thrust of the engine on the aircraft (subject to the pressure simplification indicated in the figure) is

$$thrust = m_t \times (V_j - V)$$

Here, m_t is the mass flux through the engine. Strictly, the fuel mass flux being injected into the engine should be taken into account, but this is a minor addition relative to the air mass flux.

Returning to the tube in Figure 4.6, we can use the fact that the mass flux is $m_t = (density \times area \times velocity)$, which is constant at any point along the tube, and it follows

$$m_t = density \times A_1 \times V_1 = density \times A_2 \times V_2$$

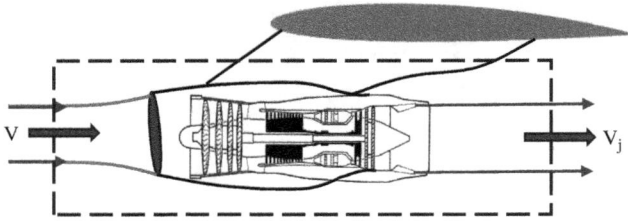

Figure 4.7.   Application of the force–momentum principle to a jet engine mounted on an aircraft wing.

This relationship can now be used to re-write the *net force* equation stated above as follows:

$$P_1 \times A_1 - P_2 \times A_2 = density \times (A_2 \times V_2 \times V_2 - A_1 \times V_1 \times V_1)$$

The reason the pressure–velocity relationship is stated in this form is two-fold. First, it brings out the fact that pressure is linked to the square of the velocity. This linkage is typical in fluid mechanics: lift, drag and losses are all linked to the square of the velocity. Second, a more rigorous analysis (involving differential calculus) shows that the following applies along any one streamline (in the absence of friction):

$$P + density/2 \times V \times V = constant$$

i.e.,

$$P_1 - P_2 = density \times (V_2 \times V_2 - V_1 \times V_1)/2$$

This is obviously not identical to the relationship we derived based on the erroneous assumption that the walls exert no force on the flow, but the essential message is the same: a pressure drop is associated with acceleration, a pressure rise is associated with deceleration and pressure changes are linked to the **square** of the velocity.

The above relationship between pressure and velocity offers the first opportunity to make the point that pressure may be regarded as a form of energy or closely connected to it. Since the kinetic energy per unit mass is $0.5 \times V \times V$, the fluid in the duct in Figure 4.6 clearly experiences an increase in kinetic energy from location "1" to location "2". The question is thus: where does this increase come from? There is no engine or another conceivable source of energy within the duct. There is no drop in thermal energy, for if the fluid is water, say, the temperature remains constant. It must be the case, therefore, that the drop in pressure is equivalent to a drop in internal energy, with this energy increment being used to increase the kinetic energy. We return to this equivalence later so as to increase the clarity of the argument. Before doing so, however, we return to the flow around our car body to make some interesting observations:

- As the flow approaches the car, it decelerates — i.e., the pressure is increasing. In fact, the pressure reaches a maximum at the "stagnation point" in Figure 4.3. This is a major contributor to the drag on the car.
- Along the front of the windscreen, the streamlines also diverge, so the pressure must rise. Hence, this process also contributes to the drag.
- On the roof, the streamlines are bunched, so the pressure there is low. This tends to create a lift force on the car, which is disadvantageous to road stability, and this is the reason for the provision of a spoiler in high-performance cars that compensates for the lift force by provoking an opposing downward force.
- Towards the rear of the car, the streamlines diverge, so the pressure is increasing. This is desirable as it causes a force opposing the drag. Similarly, below the car, towards the back, the streamlines also diverge, and the pressure is increasing, also opposing the drag. This region is occasionally referred to as a "diffuser".
- The fact that the streamline pattern in the front portion of the car is not the same as in the rear portion means that the respective pressure forces cannot cancel: the drag will always exceed the reverse pressure-induced thrust. Thus, even if the tyre and frictional drag from the car skin could be eliminated, as if by magic, an engine would be required to drive the car forward.

The comments under the last bullet point may become clearer upon our turning attention to the flow case shown in Figure 4.8. The upper sketch shows an idealised flow around a sphere. The flow is perfectly symmetrical about the vertical centreline of the sphere. Hence, the pressure fields across the front and back portions are identical, and there is zero drag on the sphere. However, in reality, the flow in the rear is very

(a)

(b)   (c)

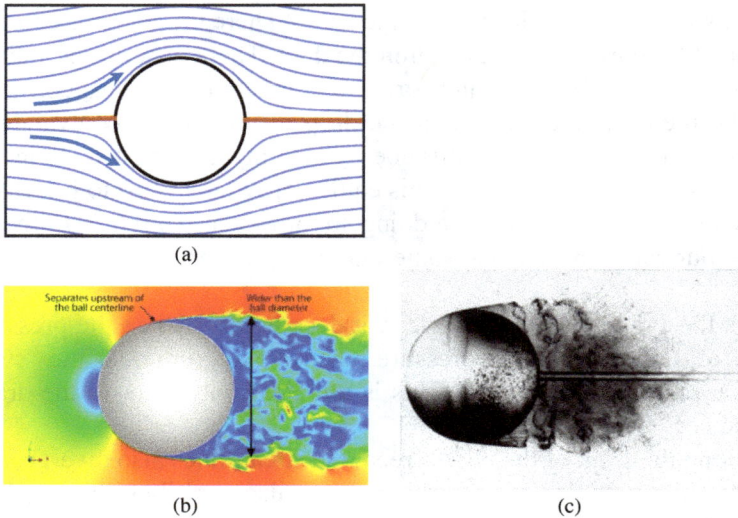

Figure 4.8. Flow around a sphere: (a) idealised (symmetric) flow and (b,c) real (asymmetric) flow around a sphere. Image (b) arises from a simulation, while (c) is an experimental visualisation.

Source: Software Cradle Hexagon (with permission); ONERA (H. Werlé).

different from that in the front, as is exemplified by the lower images in Figure 4.8. The flow "likes" being accelerated but "hates" being decelerated, and it behaves erratically when forced to do the latter. We discuss the reasons and processes involved later. Here, we merely observe that the flow behind the sphere does not follow the contour of the surface and is chaotic, in much the same way as it is in the rear of the car body. The more asymmetric the flow, the higher the drag. This is the reason why car, plane, ship and train designers have put enormous efforts into creating aerodynamically advantageous, "streamlined", shapes that avoid the chaotic wakes seen in Figure 4.8, thus minimising the drag.

The relationship between the streamlines and the pressure variations discussed above, by reference to the car body in Figure 4.3, applies equally, in principle, to the hill flow in Figure 4.4. Thus, on the windward hill side, the pressure drops as the flow accelerates up the hill from the high-pressure region close to the ground. At the top of the hill, the pressure reaches a minimum level — i.e., the hilltop is subjected to upward suction. As the flow decelerates on the leeward side and recovers, the pressure is rising again. This rise has important consequences for the flow close to the surface within the region identified by "rear-ground wake" in Figure 4.8. Here again, the flow refuses to follow the contour of the surface and chooses to recover in a manner that will become clear following the discussion in Section 4.3.

We now return to the equivalence between pressure and energy, as expressed earlier by the relationship

$$P + density/2 \times V \times V = constant$$

applicable along any streamline (in the absence of friction). In order to amplify this equivalence, we resort to the thought experiment shown schematically in Figure 4.9.

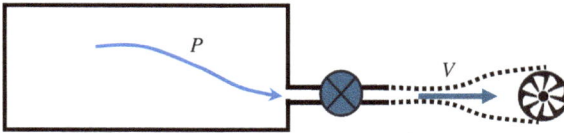

Figure 4.9. An illustration that demonstrates the equivalence of pressure energy and kinetic energy: a closed container of pressurised air is used to create a stream of speed V which then drives a turbine.

Imagine a rigid container filled with air at high pressure relative to the atmospheric pressure. The container has a valve which is initially closed. When the valve is opened, the mass of compressed air inside the container flows out at a speed V and this air has thus acquired a kinetic energy proportional to $0.5 \times V \times V$ per unit mass. In fact, going one step further, we could place a small turbine in front of the exiting stream and drive it to produce power. This shows even more clearly the correspondence between pressure and energy.

An important qualification that needs to be applied to our experiment in Figure 4.9 is that the pressure–velocity relationship, as started above, is not valid for compressible flows in which the density is not constant. As the gas flows out, its density and temperature decline with the pressure, and this leads to a more complicated relationship between energy and pressure. This interaction can be observed when the gas in a spray can is discharged through the nozzle over more than a few seconds, in which case the temperature of the can declines, possibly below the freezing point, while the ambient air is at room temperature. However, despite this complication, the experiment does illustrate that pressure is a source of energy.

We finally take the relationship between pressure energy and kinetic energy one step further to include a third form of energy: the vertical elevation of a fluid. We again resort to an experiment to clarify the relationship to follow.

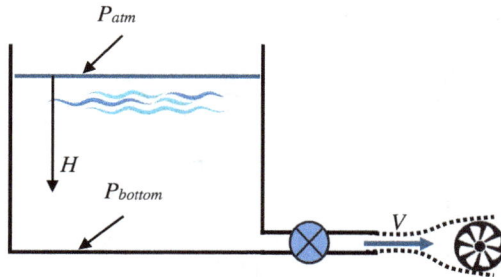

Figure 4.10. An example showing the equivalence of fluid head ("potential energy"), pressure energy and kinetic energy: an open container of water is used to create a stream of speed *V* which then drives a turbine.

The water in the open tank shown in Figure 4.10 is initially at rest, with the outflow valve closed. The pressure on the water surface is atmospheric, denoted by P_{atm}. As we "dive" down towards the bottom, the pressure increases, the increase being proportional to the depth; most will be familiar with this fact from diving into a swimming pool. The actual value of the pressure at a depth H is

$$P = P_{atm} + density \times g \times H$$

in which g is the gravitational acceleration (9.81 metres per second square, in metric units). At the bottom, the pressure is $P = P_{bottom}$, with the entire head of the water acting on the bottom. When the valve is now opened, water streams out, driven by the head of the water resting on the bottom, and the pressure is again the source of kinetic energy. This experiment shows, therefore, that the height of a fluid body above any chosen location within the fluid body gives rise to a pressure — the so-called "hydrostatic pressure" — and that this height is therefore equivalent to an energy — referred to as "potential energy".

We can now state the extended form of the previous relationship linking the pressure to the speed as follows:

$$P + density/2 \times V \times V - density \times g \times H = constant$$

This is a seminal equation in fluid mechanics, called the "Bernoulli Equation" (Daniel Bernoulli, 1700–1782). It expressed the interchangeability of pressure, kinetic and potential energy along any streamline in any flow of an incompressible fluid in which the effects of viscosity and consequent energy losses by friction are negligible.

What do we mean by "losses"? As the water streams out through the pipe and valve in Figure 4.10, the fluid experiences viscosity-related friction against the pipe wall and in the constriction in the valve — think about honey rather than water being forced out, or the outflow pipe being long and thin. This causes a loss of energy, or rather a transfer of useful energy into heat by friction, and the outflow speed will be lower than that in the absence of friction. If friction is indeed important, the "= *constant*" right-hand side of the above equation is not quite right. This then requires a straightforward correction that takes into account the losses. However, if the fluid viscosity is low, the outflow pipe is short and the valve constriction is modest, the correction will be minor.

4.3 Round-and-Round: Curvature and Vortices

A fact to repeat and emphasise, by reference to Figure 4.3, is that fluid flows along streamlines — the direction of the green and red arrows. We considered in some detail the relationship between changes in pressure and the distance between neighbouring streamlines, usually referred to as "stream tubes". More precisely, we marched along a stream tube and considered, in qualitative terms, how the pressure changes along the tube. What we do in this section is to consider the pressure variations across streamlines. The purpose of doing so is to gain an understanding of the behaviour of vortices and of curved flows around aerodynamic bodies.

Strictly, a discussion which separates pressure variations along and normal to streamlines is rather artificial, if not contentious. This is because the pressure field is continuous and varies smoothly in all directions. Thus, a pressure at any one point is connected to the pressure at all neighbouring points, and there are no distinct, separable, directions along which pressure variations act. A simple example illustrating this point is the condition prevailing in a segment of the flow around the car body taken out from Figure 4.3 and shown in Figure 4.11.

As explained earlier, the pressure drops from a high value along the blue arrow in the stream tube identified by the red broken lines. This high upstream pressure is associated with the high-pressure region around the

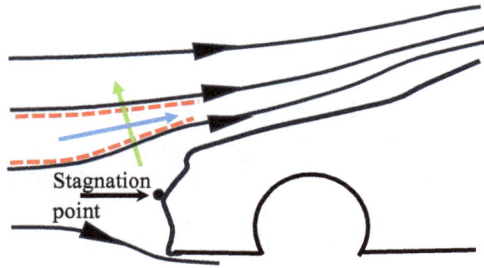

Figure 4.11. Coexistence of pressure variations along (blue arrow) and normal (green arrow) to streamlines, the former associated with acceleration and the latter with curvature.

stagnation point. In that same region, the pressure also drops from the stagnation value along the green line normal to the stream tube. This pressure drop is responsible for the curved path of the flow in the stream tube, which diverts the fluid away from the stagnation region to the outer region. Hence, the pressure "gradients" in the two directions are closely linked, as they are linked to the pressure gradient in all other directions. However, for the present purpose of understanding qualitative characteristics, we focus on pressure variations normal to streamlines separately from variations in other directions.

As with previous phenomena, we start with familiar concepts from solid mechanics. All of us will be aware of the fact that deviating laterally from a straight-line motion requires a force in the lateral direction towards the origin of the radius of curvature. The simplest model case, a mass rotating along a circular path, is shown in Figure 4.12.

Simple solid-mechanics considerations show that the radial (centrifugal) force is given by

$$F_r = m \times (V_r \times V_r/r)$$

in which the bracketed term is the "centripetal acceleration". If the mass is connected to the centre of rotation by a string — say, a child twirling a stone above its head — the radial force is the tension in the string. If the mass is a car or a motorcycle travelling along a circular curve, the radial force arises from a combination of lateral tyre friction and a radial force component supported by a positive road camber. The clearest manifestation of this force-generating process is a strong inward lean of a motorcycle or a racing push bike — a shift in the centre of gravity away from the

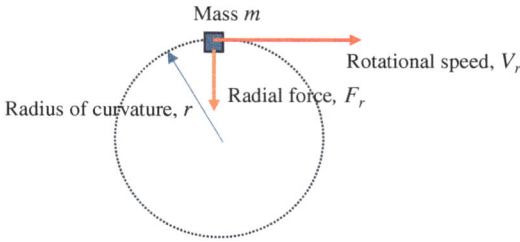

Figure 4.12. Circular rotation of a solid mass and the associated centrifugal force.

contact point with the road/track surface — when either is ridden at high speeds along a curve or a racing track.

The fluid-flow-equivalent of the rotating-mass model is a vortex. How vortices are formed is the subject of a separate description to follow in Section 4.8. Here, we merely consider the implication of the curvature and rotation for the pressure field within the vortex, the pressure being the flow-equivalent of the force in solid mechanics. However, the situation is somewhat complicated by the fact that a vortex is a rotating mass distributed over the space of the rotating body. To convey the equivalence with sufficient clarity, we consider a thin sliver of a fluid shown in Figure 4.13. The mass of the sliver is

$$m = density \times A \times d$$

The radial force on the sliver is

$$F_r = (P_1 - P_2) \times A$$

Hence, referring to the expression for F_r for the solid mass m, we get

$$(P_1 - P_2) \times A = (density \times A \times d) \times (V_r \times V_r/r)$$

or

$$(P_1 - P_2)/d = density \times (V_r \times V_r/r)$$

The left-hand side of the above equation is the radial "pressure gradient" at the particular radial location r. This shows how flow curvature is directly associated with a pressure difference normal to the flow path.

Returning to Figure 4.11, we are now able to give substance to our intuitive assumption that the upward curvature is associated with an

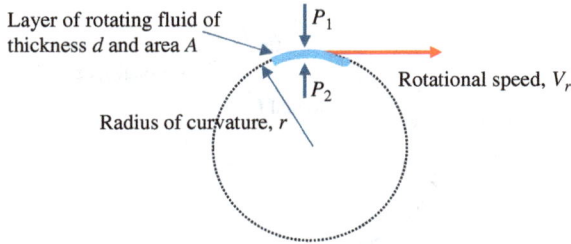

Figure 4.13.   Circular rotation of a fluid sliver and the radial pressure difference.

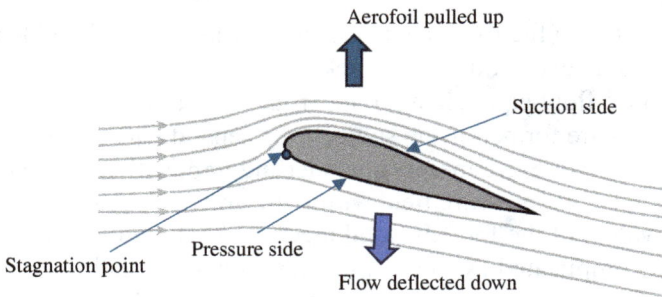

Figure 4.14.   The flow around an aerofoil in close to ideal conditions.

Source: iStock, augmented (with permission).

upward-directed pressure gradient normal to the stream tube, consistent with the high pressure in the stagnation region and lower pressure further away. Moreover, we are able to relate, more generally, this pressure gradient to the flow velocity inside a stream tube and its local radius of curvature. Thus, above the roof of the car in Figure 4.3, the streamlines curve downwards, and this implies that the pressure on the roof must be lower than the pressure above the streamlines. In other words, the car roof experiences upward suction. But we have already inferred this earlier upon observing the bunching of the streamlines above the roof, and this thus illustrates, again, the linkage between streamline-aligned and streamline-normal pressure variations.

Another flow example, central to flight aerodynamics and illustrating the usefulness of the above considerations, is the flow around an aerofoil, the cross-sectional shape of an aircraft wing, as shown in Figure 4.14. A comment made here first is that this sketch shows a rather idealised streamline pattern that might be close to reality in near-ideal cruise conditions. We shall see later that the flow around a wing can be quite complicated, especially on the upper suction side, when the angle of attack is large and "stall" sets in.

On the basis of the foregoing considerations, we can make some interesting statements on how the aerofoil operates:

- As the flow approaches the aerofoil, the high stagnation pressure diverts the flow upwards and downwards.
- The flow then accelerates strongly around the nose — the "leading edge" — as is manifested by the strong bunching of the streamlines, which indicates a strong drop in pressure.
- The flow is highly curved over much of the suction side, and this implies a pressure gradient normal to the suction side, with the pressure being low close to the surface and high further away. This creates an upward suction force, which is the most important functional feature of any wing.
- The lower stream is deflected downwards, indicating a high-pressure region on the pressure side, but the upward pressure force resulting from this region is relatively minor when compared to the upper suction force.
- The rear flow on the suction side decelerates modestly, as is suggested by the mildly divergent streamlines. This tends to increase the pressure on the suction side but only relative to the very low pressure above the nose. Hence, even though the pressure close to the surface rises slowly, it remains relatively low, and the suction effect is maintained. Indeed, this is confirmed by the curvature of the streamlines towards the rear — the "trailing edge" of the aerofoil.
- The global effect of the aerofoil is a downward deflection of the entire flow. This deflection implies a downward momentum in the rear portion of the flow relative to zero downward momentum of the incoming horizontal flow. This change in direction is thus a global indicator of the upward force on the aerofoil.
- Cleary, the streamline pattern is asymmetric, with the flow in the front very different from that in the rear, if only because of the asymmetric nature of the aerofoil itself. Therefore, based on the arguments made in relation to the car body in Figure 4.3 and the sphere in Figure 4.8, there must be a force acting horizontally on the aerofoil, which causes drag. There is an additional friction-drag component resulting from

the shear between the flow and the solid surface, but this is a subject we examine in Section 4.4.

We close this discussion on the properties of simple vortices with a few further comments relating to the last equation for the pressure gradient in the vortex. In particular, we might ask how we can use it to determine the actual pressure distribution along the radius of the vortex. To do so, we need to mathematically "integrate" the equation from the origin to any radius at which we want to find the pressure. We shall not do so, although this would allow us to discover the minimum value of the pressure in the vortex, which is (normally) at the centre of the vortex. In any event, this integration is only possible if we know how the rotation speed V_r varies with the radius — in other words, we need to know the radial velocity profile. This is no mean task because vortices can have a wide variety of velocity distributions within them. There are several classical model types that can be found in the literature. The most frequent type actually observed, certainly in the natural environment, is that shown in Figure 4.15. This distribution is typical, for example, in a hurricane (although this involves many complex aerodynamic and thermodynamic interactions): a linear-to-quadratic rise in velocity in the core and a transition to an outer region in which the velocity declines in indirect proportion to the radial distance. More about vortices (as well as hurricanes and tornados) is said later in Section 4.8.

The arguments presented so far tell us that the pressure in a vortex or curved flow decreases with decreasing radial distance from the origin of

Figure 4.15.   Radial distribution of rotational velocity typical of natural vortices such as in a hurricane.

Source: NASA.

the centre of curvature. Indeed, it was said above that the minimum pressure occurs at the centre of the vortex. However, you may have noted the qualified "normally" in brackets. One abnormal case is the anticyclonic vortex denoted by "H" in Figure 2.15. Here, oddly, we have a vortex in which the maximum pressure is in its centre. The key to unlocking this conundrum is the "Coriolis force" — a mechanism associated with the Earth's rotation, which has a strong influence on large atmospheric vortices and causes all large atmospheric and oceanic currents to flow in curved paths.

In general, rotational weather systems arise from a complex interaction among buoyancy-induced vertical flows, sea conditions, fast-moving flows in the upper atmosphere (the jet stream) and continental conditions (see Figure 2.13). Superimposed on these processes is the Coriolis force, which arises from an interaction of the rotation of the Earth with horizontal air motion. The two combine to provoke a virtual "force" that acts in a direction perpendicular to both the axis of rotation and the direction the air would flow without the Coriolis force. This force acts to deflect the air flow from its expected path in a non-rotating system. Precisely the same interaction also applies to sea and ocean currents.

To understand the Coriolis force and its role in the formation of rotating weather systems, we focus on the simplified model case shown in Figure 4.16. Imagine that you are looking down towards the Earth's North Pole, Figure 4.16(a), along the axis of rotation, which is in the anticlockwise direction. To simplify matters, but without losing the essentials of the argument, consider the Earth as a rotating disc, as shown in Figure 4.16 (parts (b) and (c)). Next, focus on the two small mass packets m shown in green and red, respectively, which rotate with the disc at the same circumferential speed V as the disc surface at the points of contact at radius r from the centre. Due to their rotating motion, these packets have an "angular momentum" $m \times V \times r$. Like ordinary (linear) momentum, angular momentum is conserved by the mass packets, unless a torque is applied to the masses.

Imagine now that the mass packets are made to move radially inwards — say, by a low pressure in the centre. They find themselves at a lower radius at which the disc surface is moving more slowly. However, the packets do not want to move at this lower speed. In fact, they speed up, rather than slow down, much like an ice skater pirouetting with arms initially outstretched and then pulling the arms inwards towards his/her body. The consequence is that the packets are deflected to the right and move along curved paths, as shown in Figure 4.16(b). Next, if the mass packets are pulled towards a low-pressure point, their deflection favours

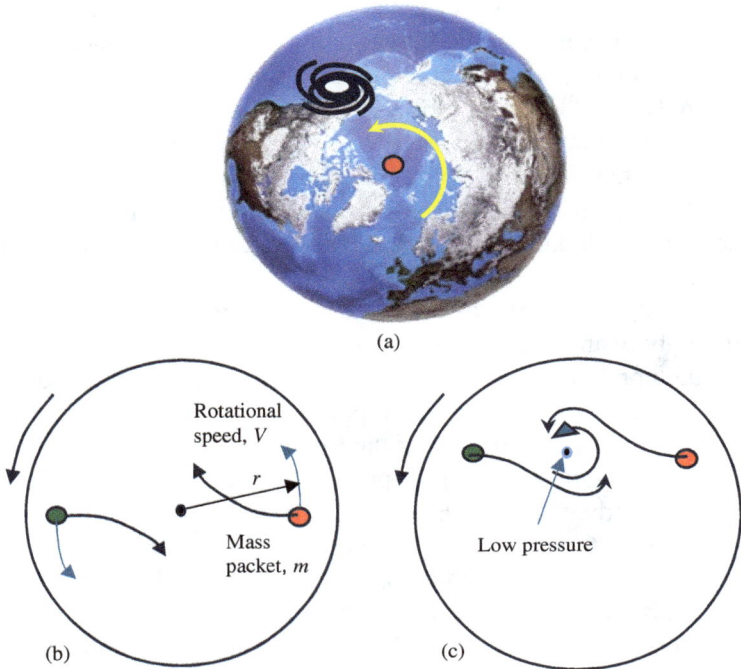

Figure 4.16.   A simplified model configuration used to explain the Coriolis "force" in the context of the Earth's rotation: (a) a view along the axis of rotation onto the North Pole; (b) a flat disk model of part (a) in which the mass packets move radially inwards and are deflected to the right; (c) the formation of an anticlockwise cyclone.

the formation of an anticlockwise vortex, as illustrated in Figure 4.16(c). This description is thus consistent with the motions in the "L" weather systems shown in Figures 2.14 and 2.15 in Section 2.5.

In the case of high-pressure regions — "H" in Figures 2.14 and 2.15 — the tendency of mass packets to curve in the clockwise direction, as shown in Figure 4.16(b), when combined with the outward motion away from high-pressure points, results in clockwise, anticyclonic, vortices around the "H" regions. However, these are benign, weak structures in which the motions are much weaker than in cyclones and in which the pressure elevation is also modest. In such conditions, the rotation-induced pressure forces are weak, and the Coriolis force driving the anticyclonic vortex dominates the behaviour of the vortex. Hence, in these exceptional circumstances, our standard considerations linking the pressure gradient to the centripetal acceleration do not apply.

Finally, if the Earth is observed from the opposite side, onto the South Pole, an entirely analogues set of interactions favours the formation of vortices in the opposite direction to that shown in the sketch — i.e., mass packets are deflected to the left, favouring the formation of strong clockwise and weak anti-clockwise vortices.

We deal with a wide variety of vortices in later sections, and we see that vortices are ubiquitous, as well as forming a hugely important building stone in numerous fluid flows. Ahead of these discussions, consider the exceptionally complex flow through a tree in windy conditions, as shown in Figure 4.17. A tree consists of thousands of branches and possibly hundreds of thousands of leaves. Every branch or leaf responds to the wind by bending, oscillating and shedding vortices that detach from either. Each also interacts with the many neighbouring branches and leaves. Here, we merely point to the formation of unsteady vortices, as shown in the right-hand-side inserts in Figure 4.17. We shall discuss various aspects of vortex formation around solid bodies in the following three sections, and we shall see, in particular, that bodies

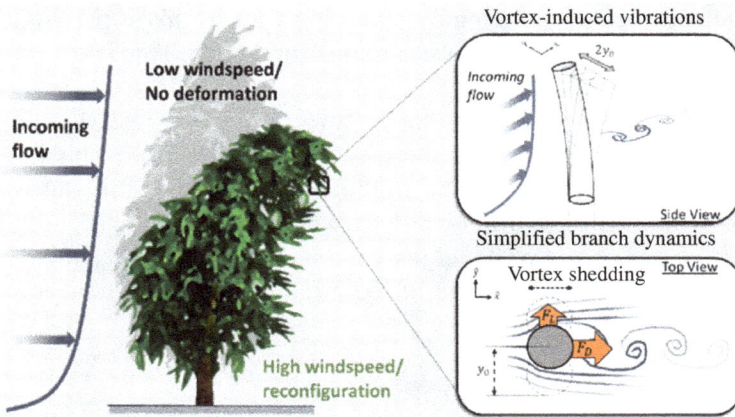

Figure 4.17. Flow past a tree in windy conditions. The flow separates behind branches and leaves, forming "streets" of vortices, each street interacting with many other streets around neighbouring branches.

Source: Dabiri *et al.* [9] (with permission).

immersed in a fluid flow tend to provoke unsteady vortices of the type shown in Figure 4.17.

4.4 Fluid Flows Dislike Deceleration (at Walls)

When we considered the flow around the sphere in Figure 4.8, it was said that *"the flow 'likes' being accelerated, but 'hates' being decelerated"*. An important qualification to add here is, however, that the verb *"hate"* applies primarily to a flow next to a solid wall. This section explains why, and what profound consequences this response has in many circumstances.

4.4.1 *Stress, strain and friction*

As a starting point, we need to consider what happens inside a fluid when it flows along a solid wall — we refer to such a flow as a "boundary layer".

Every ordinary fluid is viscous, honey and Golden Syrup evidently so, but the same applies to water, air and all other liquids and gases. An important consequence of viscosity in a fluid body, which is initially at rest and subjected to shearing by a moving wall, is that it permits a progressive transmission of the shear into the fluid body. This is shown schematically in Figure 4.18.

At the moving plate, the fluid is "sticking" to the wall. On the wall itself, the fluid moves at the same speed as the wall itself. This is a very important fact to appreciate. The argument underpinning this statement is that, if the speeds were different, there would be a sharp discontinuity in the speed between the wall and the fluid immediately at the wall. Hence, the "gradient" in the speed would be infinite, and this would imply an

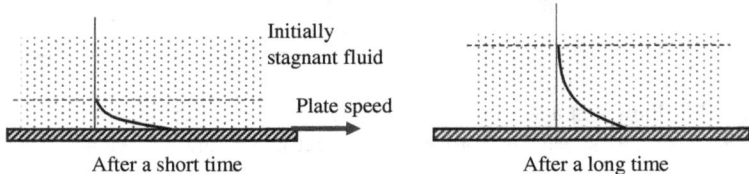

Figure 4.18. A boundary layer progressively growing normal to a wall by a shearing process.

infinite drag (shear) force at the wall. The sheared near-wall layer may be very thin, but it is never zero.

When the shearing is maintained, the sheared near-wall layer grows, with more and more of the fluid above the wall being pulled forward. This process is due to Brownian (atomic) motion, which progressively mixes lower layers with upper ones — or more accurately, the properties in these layers are mixed, one property being the flow speed. This might be easier to appreciate when the plate in Figure 4.18 is hot and the fluid body is cold. The temperature in the near-wall layer is then progressively heated by thermal conduction, with the fluid in direct contact with the wall having the same temperature as the wall itself. Conduction is also a consequence of Brownian motion, and the viscosity is analogous to thermal conductivity — i.e., the viscosity is "conducting" the property we call momentum, in the same way as thermal conductivity is responsible for the transport of heat by mixing.

A much more common boundary-layer configuration, in practice, is shown in Figure 4.19. At the plate, the fluid is stationary — again, "sticking" to the wall. Away from the plate, the fluid is sheared by the relative motion, and the sheared region progressively grows upwards as the fluid is moving along the plate. This is, essentially, what takes place along any wall past which a fluid is moving, and this includes, for example, the flow along the aerofoil surfaces in Figure 4.14 and the flow past the sphere in Figure 4.8. There is, here again, a close analogy between momentum and temperature: if the plate in Figure 4.19 were at a temperature equal to zero, and the outer stream were at a higher temperature, say 20°C, the temperature profiles in the near-wall layer would resemble closely the velocity (momentum) profiles in Figure 4.19.

One important aspect of the shearing motion is that it is accompanied by a shear force. This is fairly obvious in the case of honey or Golden

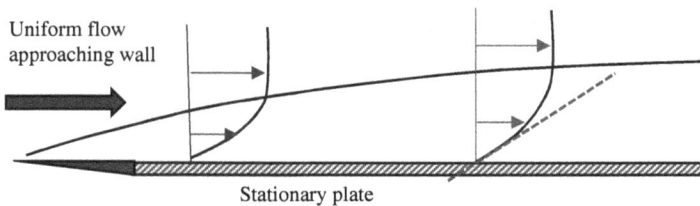

Uniform flow approaching wall

Stationary plate

Figure 4.19. A boundary layer growing along a stationary plate in an initially uniform stream.

Syrup. If a film of honey is placed between two plates, it requires a substantial force to slide one plate parallel to the other. The less viscous the fluid film is, the less force is required. Moreover, the thicker the film, the lower the force that is required. This illustrates the fact that the shear force depends not only on the viscosity but also on the rate of shear, the latter referred to as the "shear strain". This strain is the slope of the shear velocity across the boundary layer. Due to the dependence of the shear force on the shear strain, the magnitude of shear force between two adjacent fluid layers in Figures 4.18 and 4.19 depends on the distance normal to the wall. In most fluids familiar to us — air, oil, water, most gases and liquid metals — the shear force is linearly related to the viscosity and the shear strain. Blood, ketchup, cream and liquid chocolate are examples in which the relationship is not linear because the viscosity of these substances is not constant but depends on the shear strain.

A shear force of particular interest is that on the solid wall, referred to as the "friction-drag force" or "skin-friction force". The strain rate — the rate of shear motion — at the wall is represented by the gradient of the velocity profile at this position, the dashed line in Figure 4.19. In fluids having a constant viscosity, the shear force per unit area on the plate at the streamwise location in question is

shear force (per unit area) = viscosity × (velocity gradient at the wall)

The total friction-drag force on any surface is, therefore, the sum (integral) of all the force components acting on the respective surface patches constituting the whole surface.

If a wall is curved, the shear force acts tangentially to the wall, and the relevant velocity gradient in the above relationship is the one normal to the wall. If we then wish to find the total shear force (the friction-drag force) in a particular direction — say, the horizontal direction — we need to determine the local shear forces (per unit area) at all locations on the wall and then sum-up all the horizontal components of the local values. This would apply, for example, in the aerofoil in Figure 4.14, if we wanted to find the total friction drag in the direction of flight.

4.4.2 *Separation anxiety*

When a boundary layer is subjected to rising pressure in the direction of the flow, its structure undergoes changes that can have a profound effect

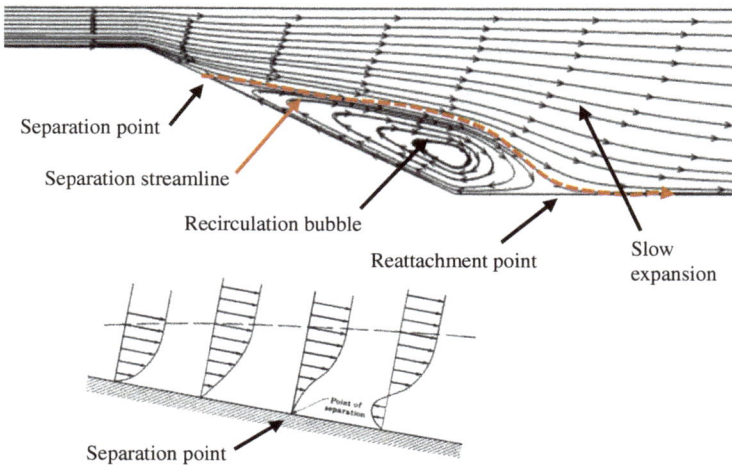

Figure 4.20. Flow in an expanding diffuser in which the boundary layer separates, and the structural response of the boundary layer to the rising pressure in the diffuser.

Source: Mariotti *et al.* [10] (with permission).

on the behaviour of countless environmental and engineering flows and on the operational performance of the latter flow group. Here, we consider the basic interactions, in general, before extending the discussion to specific flows.

A simple model configuration that serves the present purpose is the flow in an expanding duct, referred to as a "diffuser". The lower half of such a symmetric diffuser is shown in Figure 4.20. At this stage, we shall assume that this flow, as well as others discussed later in this section, is steady — i.e., we put aside the possibility that the flows are affected by wavy instabilities, oscillations and turbulence. The consequences of unsteadiness will be considered in Section 4.6 and Chapter 5.

As the flow enters the diffuser, the streamlines diverge because the fluid has to fill the expanding flow area. We know now, from Section 4.2, that diverging streamlines imply a rising pressure. In fact, diffusers are often used in engineering applications to lower the flow velocity and increase the pressure — although the diffuser shown in Figure 4.20 is poorly configured from an operational perspective, as will transpire shortly. Here, the most important question to address is how the wall boundary layer reacts to the rising pressure and whether it can overcome the resistance it encounters. Superficially, this is akin to a pedal cyclist having to work hard to climb a hill. If the flow has enough momentum, it will succeed in overcoming the resistance, but if its momentum is too low, the fluid will stop and flow back, much like a cyclist running out of energy and being driven back down the hill.

Figure 4.19 shows that the shearing effect of the wall on the fluid creates a region of low velocity — i.e., low momentum — close to the surface. This flow region of low momentum is least able to resist the rising pressure, and it slows down much more rapidly than the high-momentum region further outside. There comes a point at which the flow closest to the surface comes to a halt, and reverse flow begins to set in, as shown in the right-hand part of Figure 4.20. This location is called the "separation point" because at that point the boundary-layer fluid effectively detaches from the surface, is being displaced upwards and is being replaced by the reverse flow which is driven by the adverse streamwise pressure rise.

It is important to point out here that the reference to a "vacuum" in the area occupied by the recirculation bubble, occasionally so asserted, is entirely incorrect. Rather, the recirculation bubble is filled with slowly swirling fluid. If any fluid is dispaced, for whatever reason, other fluid has to take its place, and in this case, the fluid comes from the right, the reverse flow near the wall. What is correct to state is that the pressure in the centre of the bubble is lower than further outside — i.e., there is no "vacuum". This is consistent with the arguments made in Section 4.3, namely, that the curvature of the streamlines implies a suction towards the centre of the vortex or the origin of the radius of curvature.

As is shown in the upper part of Figure 4.20, once the flow progresses well beyond the diffusing section, where the pressure no longer rises, it recovers and eventually settles down to a normal duct flow — although this can be a slow process, taking several duct heights. The separated boundary-layer fluid returns to the wall, and the forward flow resumes, indicated by the broken red line. Fluid below the separated boundary layer (just below the red line) is being dragged forward by viscous friction, moves towards the wall just to the left of the "reattachment point" and then flows back along the wall towards the separation point. All these processes combine to form a recirculation bubble in which fluid is continuously fed backwards to replace the separated boundary layer which is pushed upwards away from the wall.

An important consequence of the recirculation bubble shown in Figure 4.20 — often wrongly referred to as a "dead-water region" — is that it moderates the pressure rise that the actual geometry intends to impose on the flow. In effect, the flow entering the diffuser is contained between the upper wall and the upper boundary of the separation bubble, roughly the red streamline. We know from the explanations in Section 4.1 that no mass crosses a streamline. It follows that the red line can be viewed as a virtual "wall", in so far as it confines the flow moving along the expanding duct from the inlet to the outlet. As is evident from the lower rate of divergence of the streamlines above the recirculation bubble, the flow slows down at a lower rate than the actual lower diffuser wall attempts to impose on the flow, and this implies a lower pressure rise.

The flow shown in Figure 4.21 conveys the message that separation is not conditional on confinement. This flow is the same as that shown in Figure 4.4, but here, the obscured region on the leeward side is uncovered to reveal the possible presence of a recirculation bubble that is provoked by the rising pressure acting on the hill-surface boundary layer. The adjective "possible" is added above because the presence or absence of separation depends on the slope of the leeward side and the ability of the flow to resist separation though an advantageous near-wall momentum distribution. In the presence of separation, the flow within the recirculation bubble is slow and is moving uphill close to the ground. This is the region that offers hill climbers maximum protection from the wind in stormy conditions.

As in the case of the diffuser flow in Figure 4.20, the recirculation bubble reduces the rate of deceleration and flow divergence. Also, as before, the boundary of the recirculation bubble — the red-coloured "separation streamline" — can be thought of as a virtual mass-flow barrier through which no mass is exchanged between the recirculation bubble and

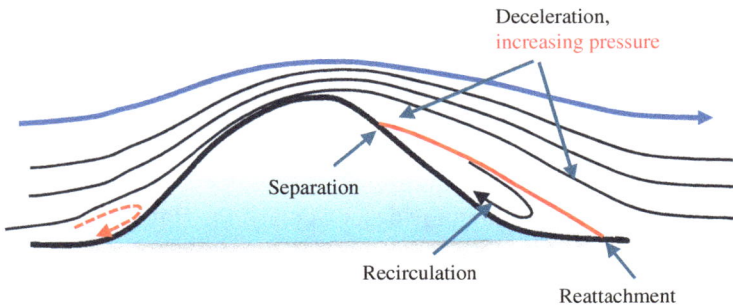

Figure 4.21. Flow around an idealised hill, with the covered area in Figure 4.4 now identified as containing a recirculation region.

the outer flow. Hence, the mass flux upstream of the hill, contained between the ground and the blue streamline in the outer region, is the same as that between the blue streamline and the red separation streamline behind the hill and also between the blue streamline and the ground behind the reattachment point.

The possibility of the presence of a second recirculation zone in the upstream corner of the hill is indicated, tentatively, in Figure 4.21 by the broken red line. Separation in this region may be provoked by the rising pressure associated with the obstructing influence of the upward-curving hill side and acting on the near-ground boundary layer approaching the hill. Even if this separation does occur, it is highly localised and much smaller than the leeward separation.

The ability of the flow to resist separation depends on the rate at which momentum can be transferred — colloquially, "pumped in" — from outer fast-flowing regions to slow near-wall regions. The faster this momentum is transferred, the more effectively the boundary is energised and able to resist separation. The cross-flow transfer of momentum occurs by the action of viscosity — a mixing process similar to heat transfer by conduction, as pointed out already. However, the viscosity of air is quite low, so this process is not very effective. It is argued later, in Chapter 5, that a much more effective transfer is achieved by the action of turbulent mixing in the boundary layer — an eddying process that redistributes momentum (and heat as well as species) much more effectively than viscosity.

4.5 Aircraft Wings Hate Separation: They Stall

In the context of flight, "stall" is a process by which the aircraft loses lift and thus its ability to stay aloft. Stall is an especially dreaded term in the vocabulary of those who have a fear of flying but is, more generally, also a serious matter for all who fly, pilots included. Stall arises when a wing is inclined relative to the flight direction at an excessive angle of attack, as is superficially shown in Figure 4.22.

One feature highlighted in the description of the flow around the aerofoil in Figure 4.14 is that the streamlines over the rear part of the suction side are slowly diverging, thus indicating that the pressure was slowly rising along the flow, though still remaining relatively low, and thus keeping the upward suction force effective. So, what is happening on the suction side in Figure 4.23?

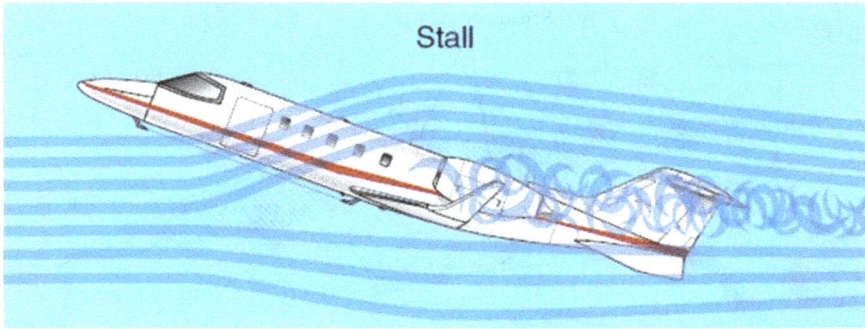

Figure 4.22. A schematic of stall from an aircraft wing in an excessive angle of attack.
Source: Wikimedia Commons CC-BY.

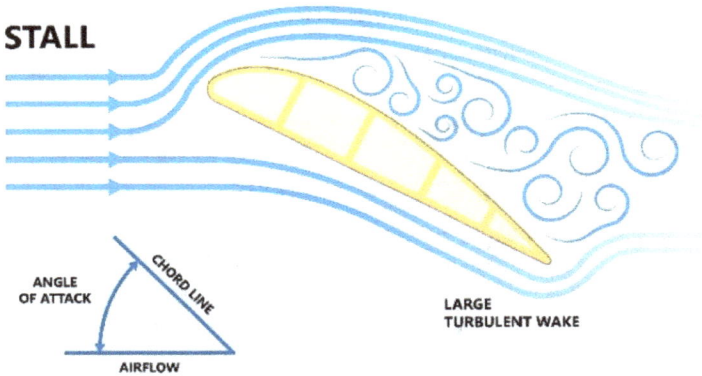

Figure 4.23. Schematic of an ill-conditioned flow over the suction side of an aerofoil at an excessive angle of attack.

Source: Shutterstock (with permission).

When the incidence of the aerofoil is excessive, the flow cannot tolerate the rapidly increasing area between the suction side and the outer undisturbed flow, for this would require a strong deceleration and thus a steep pressure rise. The flow is unable, therefore, to follow the curved portion of the aerofoil, as is shown in Figure 4.24, in the same way as the flow in the diffuser in Figure 4.20 or on the leeward side of the hill in Figure 4.21 refused to follow the wall contour. The streamlines are no longer curving downwards, and the consequence is that much of the suction-produced lift is lost. In extreme conditions, visualised in Figure 4.25

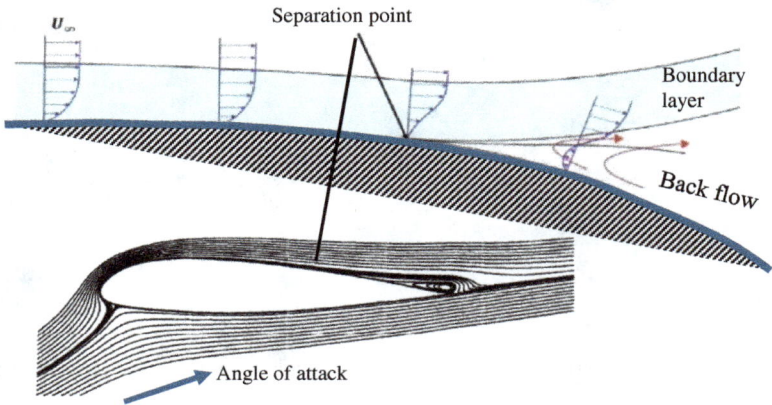

Figure 4.24. The process of boundary-layer separation from a curved surface due to unfavourable pressure gradient and from the suction side of an aerofoil at high incidence.
Source: Leschziner [11] (with permission); Ghouila-Houri *et al.* [12] (with permission).

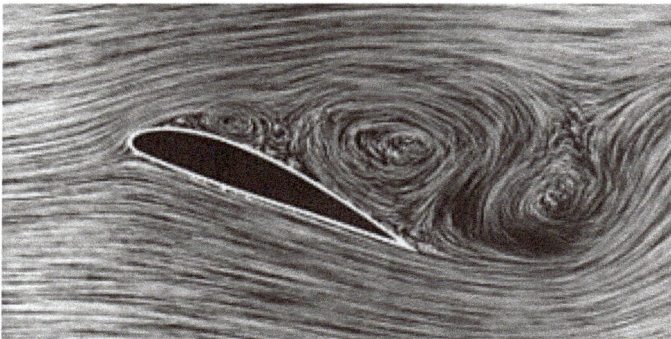

Figure 4.25. Massive separation from, and stall of, an aerofoil at very high incidence. Separation starts at the trailing edge and progresses towards the leading edge to encompass the entire suction side, resulting in a large recirculation bubble above the suction side.
Source: DLR, Wikipedia CC-BY.

by means of a computational simulation, the separation is massive, covering the entire suction side, in which case there is very little lift generated on the suction side.

Although the above arguments point unambiguously to the seriously adverse effects of a vortex that result from separation on the suction side, there are circumstances in which vortices have the opposite effect and

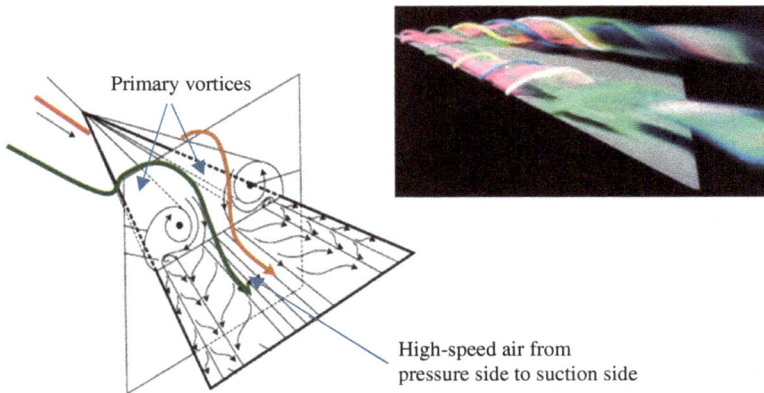

Figure 4.26. The leading-edge vortex structure over the suction side of a "delta" wing at high incidence.

Source: Aviation Stack Exchange.

increase the lift. Such circumstances arise in some highly three-dimensional wing configurations and rely on the formation of a strong vortex by separation from the leading edge. A representative model configuration is a "delta" wing at high incidence, as shown in Figure 4.26, the *Concorde* supersonic-aircraft wing being a practical realisation (in landing and take-off) of the fluid dynamics involved. Here, three-dimensional separation occurs on the sharp or tightly curved swept edges of the wing, and the flow is pulled towards the suction side, curving downwards and giving rise to intense vortices that are anchored on the edges. Apart from aiding the lift by the low pressure within them, the vortices induce fast-moving air from above the wing towards the suction side, thus allowing the wing to maintain lift in the face of high incidence relative to the oncoming flow.

Bird flight is another example in which suction-side vortices are advantageous. In advance of the description that follows, it has to be acknowledged first that bird flight is extremely complex and varied — much more complex than flight with man-made aircraft. Each bird (and insect) species has evolved its own "optimal" physiological (wing shapes control, muscles, feather structures…) and aerodynamic characteristics, which also vary greatly with the type of manoeuvres a bird executes — i.e., take-off, landing, migratory cruise flight, etc. The only process outlined here, in very simple terms, is the manner in which birds exploit vortices above their wings.

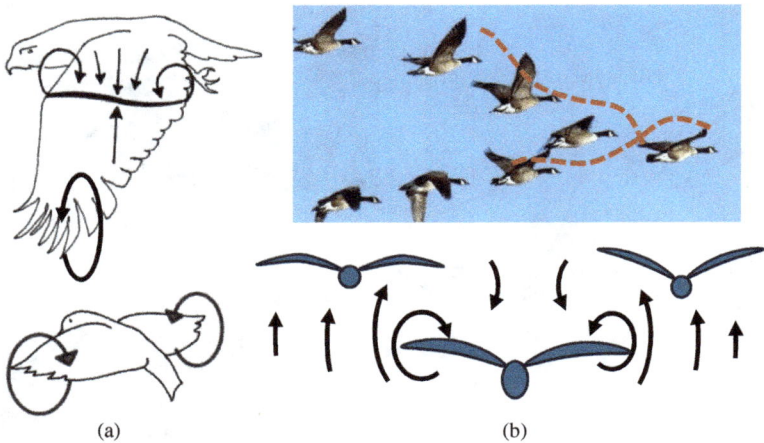

Figure 4.27.   The flight of birds: (a) schematic of the vortices formed on the suction side of bird wings during the downstroke phase; (b) the exploitation of the upwash of wing-tip vortices for increasing lift by birds flying in V formation.

Source: Dvořák [13] (with permission); iStock (with permission).

The essential interactions are conveyed in Figure 4.27(a) for birds with long (high-aspect-ratio) and short (low-aspect-ratio) wings, respectively.

With the exception of the hummingbird, birds generate lift and thrust by flapping and flexing their wings. In contrast, the hummingbird and insects oscillate their wings, rotating them along the wing axis during the downstroke and upstroke. When a bird wants to create lift and thrust, it pushes down its wings, increasing substantially the angle of attack of the wing relative to the flight direction. The downstroke generates thrust on the lower (pressure) side and also lift on the upper (suction) side. This pressure difference induces a flow across the wing edges, including the tip region, from the lower side to the upper side, giving rise to the vortices shown in Figure 4.27(a). As in the case of the delta wing, shown in Figure 4.26, the vortices formed following separation, especially from the leading edge, induce fast-moving air to be drawn downwards towards the suction side, thus avoiding reverse-flow over much of the suction side and maintaining lift, despite the high angle of attack. Furthermore, the downward momentum of the air drawn towards the suction side induces an upward force on the wing that enhances the lift. During the upstroke, the bird configures its wings to operate at a low angle of attack, resulting in a close-to-zero or slightly negative lift and low drag. In the final stage, the vortices shown in Figure 4.27(a) are swept downstream before the next downstroke. The entire process is referred to as "dynamic stall". The lift and drag on bird wings are thus highly time-dependent, varying in harmony with the wing motion — though usually with a lag, called "hysteresis".

An intriguing group behaviour, observed especially in long-distance flight of large birds (e.g., geese and ibises), is that the flock fly in an organised 'V' formation, as shown in Figure 4.27(b). One reason is that this pattern allows birds flying in the wake of those ahead to increase their lift and thus save energy. There is much uncertainty, however, about how much energy is saved by the trailing birds. Figures reported in the scientific literature vary between 10% and 70%, with a level around 40% being regarded as a reasonable estimate for Canada Geese. Another (somewhat speculative) reason for the "V" formation is that it facilitates communication within the flock and a coordinated evasive action in response to hazards.

The basic aerodynamic mechanism at play is shown in the lower part of Figure 4.27(b). The lateral displacement of the two birds flying behind the leader places them within the upwash induced by the leader's wing-tip vortices. These vortices are shed backwards, at an intensity that varies in harmony with the flapping motion, and they persist behind the bird, weakening with increasing distance (we shall return to the subject of wing-tip vortices, specifically in the context of fixed-wing aircraft, in Section 5.2 by reference to Figure 5.2). The birds flying further behind within the flock then capture the combined upwash. However, the aerodynamics are substantially complicated by the fact the strength of the wing-tip vortices varies substantially in time because it is closely linked to the unsteady lift of the flapping wings. In addition, the trajectory of the vortices meanders in space, as indicated schematically by the red broken lines in Figure 4.27(b). Thus, in order to maximise the upwash capture, the birds following the leader have to flap their wings in phase with those of the respective leader, the phase shift being dependent on the flapping frequency and the spatial separation in the direction of flight.

In an engineering context, the ability of a flow to resist separation can be enhanced by various geometric and aerodynamic tricks that control the flow. This is in addition to promoting cross-flow mixing by turbulence in the boundary layer — a topic that is discussed in Chapter 5. Geometric modification can take many forms, some very subtle and involving a variety of small aerodynamic devices, appendages, slots and deflectors, some discussed in Chapter 7.

One modification that is familiar to most fliers is the extension of slats and flaps during landing — a noisy deployment for those sitting close to the wings, second only to the extension of the landing gear. In this phase, the aircraft slows down, and the lift declines. To increase the lift, the pilot can increase the angle of attack, but this increases the drag, causing excessively rapid descent. More seriously, and relevant to the present discussion, it can easily lead to stall. The solution is to give the aircraft more wing area and also to condition the flow that is subjected to deceleration due to the adverse pressure gradient on the suction side. This is explained by reference to Figure 4.28.

The details of the aerodynamic interactions among the three wing elements shown in Figure 4.28 are multifaceted, and the following explanation is a simplified version of reality. This interaction occurs in the forward as well as backward directions, and it also involves effects in the spanwise direction — i.e., across the wing-root-to-tip span — that cannot be explained by reference to the two-dimensional cross-section shown in the left part of Figure 4.28.

The slat — the small front wing — makes a fairly minor direct contribution to the total lift. Its main function is to control the air flow ahead

Figure 4.28.   Multi-element aerofoils and wings used to enhance lift and avoid stall. The right photo is of the Airbus A310 aircraft.

Source: Sereez and Zaffar [14] (with permission); Wikipedia CC-BY.

of the main aerofoil. It redirects the flow around the leading edge of the main element, both through its suction side and the flow in the slot between the two aerofoils, reducing the curvature of the flow over the leading edge and the effective incidence angle between the flow and the main aerofoil. Although the slat generates a low-velocity wake behind it, this wake is pushed well away from the main element, and a new thin boundary layer develops on the main wing, starting at its leading edge; recall here that a thin boundary layer is better able to withstand the damaging effect of an adverse pressure gradient than a thick boundary layer.

The flap is a much more important direct contributor to the lift than the slat. This device is also deployed during take-off, although at a lower extension distance than during landing. In effect, it functions as a new wing, separate from the main element, though interacting aerodynamically with it. It can be twisted down at an astonishingly large angle relative to the direction of flight because the flow on the rear of the main element is already deflected through a substantial angle before it interacts with the flap. The high-momentum stream in the slot between the main aerofoil and the flap pushes away, in the upward direction, the low-momentum wake created by the suction-side boundary layer of the main element, and this favours the formation of a new thin boundary layer on the flap. The extended flap thus operates very differently from its retracted form when it functions simply as a continuous extension of the main element.

We shall discuss a number of applications that involve other wing-like geometries later, in Chapter 6, including jet-engine compressor and turbine blades, wind-turbine blades, propellers and helicopter rotors.

4.6 A Car Is (or Can Be) a Wing

We return to the description of the flow around the car shown in Figure 4.3, Section 4.1. Among a number of flow features discussed, the curved streamlines over the roof were highlighted, and it was argued that this implied a lift force that could destabilise the road holding of the car. The fact that the flow on the rear window remains attached and continues on a curved downward path only increases the lift force further. We now know from Section 4.4, however, that the pressure rise implied by the divergence of the streamlines above the rear window can lead to separation and loss of lift if the rear window is inclined downwards beyond a certain angle — around 30°. This separation can be promoted by including an upward-curving lip along the edge linking the roof with the rear window. In effect, this lip acts as a mini-spoiler, and this is an example of what was earlier referred to as "tricks" designed to control flows. On the other hand, we may wish to avoid separation at this position because a recirculation region along the rear window is likely to promote the deposition of dirt by transporting dirty air lifted from the road from the back of the car onto the window.

The lift generated on car bodies is especially worrisome in high-performance and racing cars, which are highly streamlined and on which the lift force is especially high due to the fact that the lift tends to rise quadratically with the speed. In effect, the car can become a "wing on wheels", with the contact between the wheels and the road being almost lost.

To overcome the adverse consequences of the lift force, various types of spoilers are often fitted to cars, an example of which was already mentioned in the first paragraph above. One type of spoiler and its aerodynamic effects are shown in the upper part of Figure 4.29. This spoiler may be small, but it has the effect of significantly reducing the lift by creating a high-pressure region upstream of the spoiler, causing the flow to separate and diverting it upwards. It might even reduce the drag by reducing the horizontal component of the suction on the rear part of the roof and modifying the recirculating wake behind it. On the other hand, the spoiler itself causes added drag because it obstructs the flow and creates turbulence, both in the highly turbulent separated zone upstream of the spoiler and the wake behind it. Turbulence causes useful mechanical energy to be converted into useless heat by friction, and this is one reason for an increase in drag. Hence, in respect of drag, the balance between gain and loss is tricky.

To increase the downforce further, wing-like spoilers are used — their extreme application being in F1 racing cars, as shown in the lowest image of Figure 4.29. In essence, the spoilers consist of sets of inverted wings, with suction and pressure sides reversed, which generate down forces by downward suction, thus keeping the car in contact with the racing track. As seen in Figure 4.29, inverted wing-like appendages are also fitted to the front of the F1 car, again designed to increase the downward force. The front spoiler is also fitted with vertical "winglets" whose function is to reduce the drag created by the main spoiler aerofoils, to enhance lateral stability and to control the direction of the flow and the vortices that emanate from the spoiler around the car.

It is finally appropriate to note here that there are as many solutions to car aerodynamics as there are manufacturers, car models and

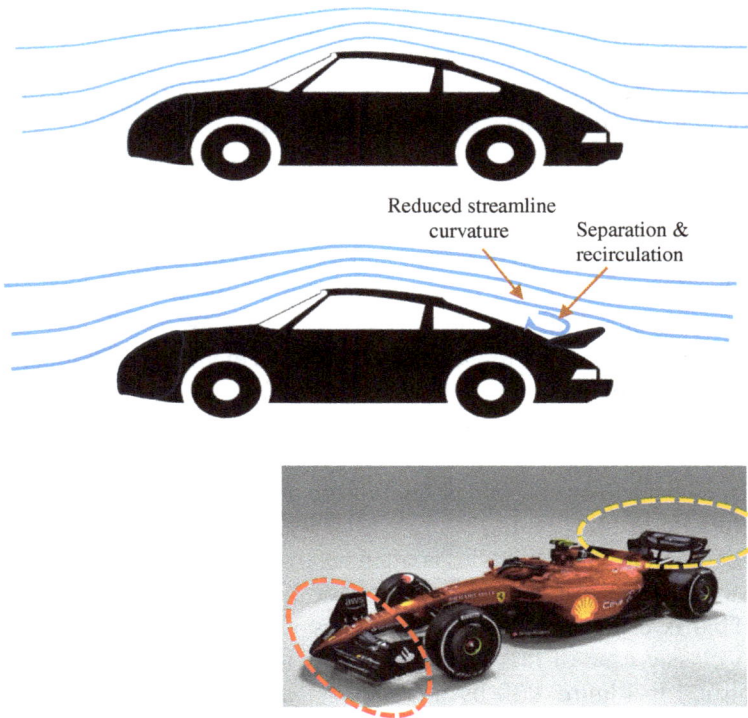

Reduced streamline curvature

Separation & recirculation

Figure 4.29. Two applications of car spoilers to reduce upward suction lift and/or create downward forces.

Source: Shutterstock (with permission).

racing teams. This points to the fact that this branch of aerodynamics — and it is by no means alone — is as much an art as it is a science and a matter trial-and-error as well as aesthetics. The aerodynamic effect of any appendage to a car body, mirrors included, depends strongly on the precise shape of the car, on tyre shape and rotation, on any other components fitted to the car body and on the flow to which the car is subjected, side-wind in particular.

4.7 The Beastly Face of Bulky Bodies and Nasty Edges

The distinction between streamlined bodies and bulky "bluff" bodies is not always clear-cut. A case in point is the ball shown in Figure 4.8.

This has characteristics of both types. Since its surface is continuous and smoothly varying, the flow around it can be described in terms very similar to those we applied to wings at high incidence. Specifically, we observe that the streamlines diverge downstream of the centre of the ball as the flow attempts to fill the expanding space behind the ball. However, this expansion is much too severe for the boundary layer developing on the front portion of the ball to remain attached, and the flow separates as the pressure increases, giving rise to a large recirculating wake that moderates the flow recovery far downstream of the ball, in the same sense as the recirculation bubble does in the diffuser of Figure 4.20.

The car shown in Figure 4.3 also combines elements of both streamlined and bluff bodies, although in a different sense to that of the ball. Here, most of the car surface is streamlined, but the front and rear are not. In particular, the rear portion of the car, behind the boot, involves sharp changes in surface orientation at the bottom and top, and such changes are characteristic of bluff bodies.

Separation from bluff bodies is often unsteady — i.e., time-dependent — unless the flow velocity is extremely low or the fluid viscosity is extremely high. This is especially so in long bodies with constant cross-sectional area. A classic case of this type is the flow around a very long cylinder, exemplified in Figure 4.30 by a smoke-release experiment (look also at the insets in Figure 4.17). Here, in contrast to the three-dimensional ball, the flow separates alternately and periodically from the upper and lower parts of the surface to form a street of vortices, called the "von Karman

Figure 4.30. Periodic separation from the lower and upper parts of a long cylinder, forming a "street" of vortices.

Source: Griffin and Ramberg [15] (with permission).

vortex street". This periodic separation is a consequence of a subtle instability between the upper and lower components of the flow when these components originate from a single stream — i.e., when the flow domain is doubly connected so that pressure perturbations can propagate freely in both upstream and downstream directions between the upper and lower parts of the flow. However, unsteady separation of this type also tends to occur in many other geometries having elongated profiles and bluff rear surfaces, in which case two streams developing over both sides of the body separate and interact in a downstream wake. This unsteadiness can only be suppressed by careful control, or by lowering the approach velocity, or by introducing targeted geometric modifications. For example, if a thin splitter plate is inserted, as suggested by the orange line behind the cylinder in Figure 4.30, the communication between the upper and lower regions is substantially weakened, and the instability is suppressed.

The periodic separation also causes periodic fluctuations in the pressure field, which then give rise to force fluctuations and provoke up/down oscillatory movements in the cylinder. We do not normally see these movements, but we hear them. On a windy day, the flow past long bodies having a uniform cross-section, such as long electricity cables or antennae, creates buzzing or whistling, and this is an audial manifestation of the oscillations. The same arises on very tall buildings and chimneys, but in these cases, the frequency of the separation and thus of the induced oscillations is much lower because this frequency reduces in direct proportion to the diameter of the cylinder.

Flow-induced oscillations can be quite violent in tall structures and must be avoided. There is an especially famous case that occurred in 1965, in which three out of a group of eight cooling towers at the Ferrybridge "C" Power Station in the UK collapsed because of wind-induced oscillations, with the remaining towers sustaining severe structural damage. As is exemplified in Figure 4.31, flow-induced vibrations on tall structures can be avoided relatively easily by changing the shape of the building or chimney along its length to avoid long stretches of uniform cross-sectional area. These changes induce variations in the flow along the axes of the structures and give rise to axial flow components that disrupt the two-dimensional mechanisms responsible for the periodic shedding.

Sharp edges and corners are characteristic geometric features of many bluff bodies, and these almost invariably provoke separation, the

Figure 4.31. Avoidance of flow-induced oscillations in tall structures by introducing variations in the geometry along the objects.

Source: iStock (with permission); Wikipedia CC-BY.

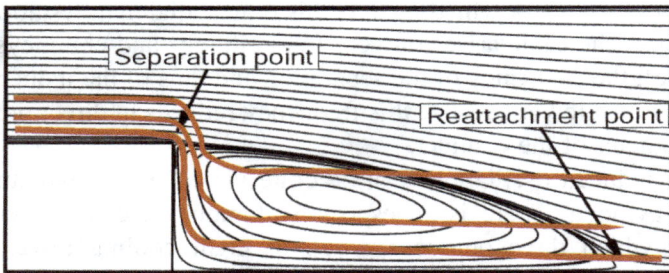

Figure 4.32. Separation and recirculation behind a step with a sharp (or highly curved) edge. The red lines indicate the "expected" streamlines if the recirculation zone were to be disallowed.

Source: M. Breuer (private communication); Biswas *et al.* [16] (with permission).

ensuing recirculation zone often occupying a substantial proportion of the flow area. A simple configuration is the flow past a step, as shown in Figure 4.32. Avoidance of the separation from the corner — essentially a curved surface of very small radius of curvature — would require a near-discontinuous expansion of the flow and a huge rise in pressure for the flow to follow the step contour, as suggested by the red lines in

Figure 4.32. The flow cannot do so and separates at the step. As in earlier cases — e.g., the diffuser in Figure 4.20 — a recirculation zone is created that allows the flow to expand and slow down much more gradually towards the larger downstream flow area.

The fluid which separates at the corner follows the "separation streamline" emanating from the edge and reaching the lower wall at the reattachment point. Here, the fluid splits: the fluid above the separation streamline is flowing forward, and the fluid below the line is flowing backwards. This behaviour is very similar to that described by reference to the diffuser in Figure 4.20.

The flow inside the recirculation zone is moving slowly. Anyone trying to shelter from a strong wind will be aware that moving behind an obstacle into a corner reduces substantially the discomfort. Another familiar experience is that leaves, paper and other loose debris tend to accumulate in corners. This is due to the slow reverse flow carrying these objects backwards and depositing them in the corner.

The sharp-edged bluff-body equivalent to the circular cylinder is the square cylinder shown in Figure 4.33. The figure contains two instantaneous snapshots derived from a computational simulation at two different speeds, the left figure being for the lower speed. A feature common to both circular and square cylinder flows is that both separate alternately from the lower and upper edges in a periodic fashion, giving rise to an unsteady street of vortices in the wake. In fact, if the diameter of the cylinder and the height of the square are the same, the frequency is the same.

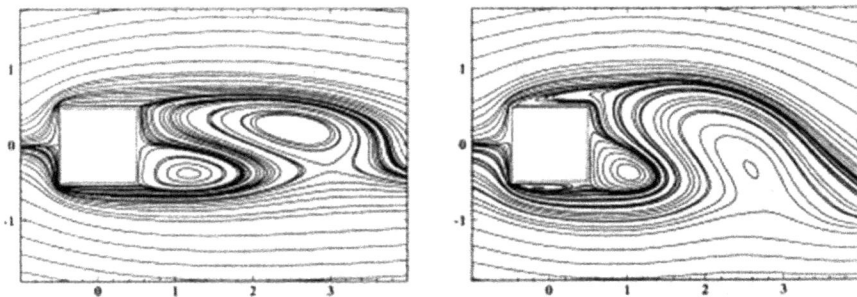

Figure 4.33. Periodic separation from the lower and upper corners of a long square cylinder, forming a "street" of vortices. The right-hand-side image is at a higher speed than the left-hand-side image.

Source: Breuer *et al.* [17] (with permission).

However, the flow at the higher speed in Figure 4.33 shows that the separation behaviour is more complex in the square-cylinder case: the flow first separates from the left-hand-side corners, it then reattaches intermittently on the upper and lower faces, and it finally fully separates from the right-hand corners.

In engineering applications, as well as environmental problems requiring human intervention, a description of the unsteady flow is often not very informative or useful because there are an infinite number of different instantaneous states in the flow. Engineers and designers are thus primarily interested in the long-time-averaged conditions of the flow. In principle, the average can be obtained by summing up a large number of instantaneous fields covering many shedding cycles and dividing the result by the number of fields recorded. This is called "ensemble averaging", and it is equivalent to time-averaging over a long period, again covering a large number of shedding cycles. If this approach is applied to our round- and square-cylinder flows, the results are typically of the form shown in Figure 4.34. It needs to be underlined here that these average pictures are synthetic, a virtual representation that arises from taking many snapshots and superimposing them on top of one another. The benefit is that one gets an average representation — an "engineering" view of the behaviour and properties of the flow. Such a view is also useful if comparisons are to be made with flows which are not dominated by unsteady shedding and which do not require time-averaging.

With a few exceptions, separation phenomena so far discussed in any detail are essentially two-dimensional in character — i.e., the flow is assumed to be identical in any plane normal to the page. This might appear especially odd in the case of the car body in Figure 4.3, but it is possible to conceive of a model that has a constant cross-section

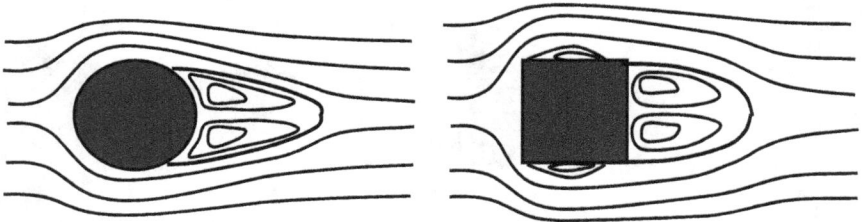

Figure 4.34. Average (virtual) flow fields for the round and square cylinders derived from superimposing numerous unsteady fields of the type shown in Figures 4.30 and 4.33, respectively.

corresponding to the real car's centre plane extruded in the direction normal to the page.

While two-dimensional separation is attractively simple and allows many of the important underlying phenomena to be elucidated, separation from three-dimensional bodies is the rule in the real world, and its structure is more complicated. There is a huge number of situations in nature and engineering in which a wide variety of three-dimensional separation is a central ingredient of the flows in question. Three examples are shown in Figure 4.37. In all three cases, separation may be steady or varying with time; a great deal depends on the precise geometric and flow conditions. Here, we assume that the flows are steady.

The first two examples in Figure 4.35, parts (a)–(c), are three-dimensional versions of the square-cylinder flow shown in Figure 4.34 (it may be helpful to imagine the latter as being a cross-section of the three-dimensional flow in Figure 4.35(c)). As before, separation occurs from the side corners, and there is a large recirculating wake behind the bodies. However, there are important additional features which do not occur in the two-dimensional flow. Here, separation also occurs on the top

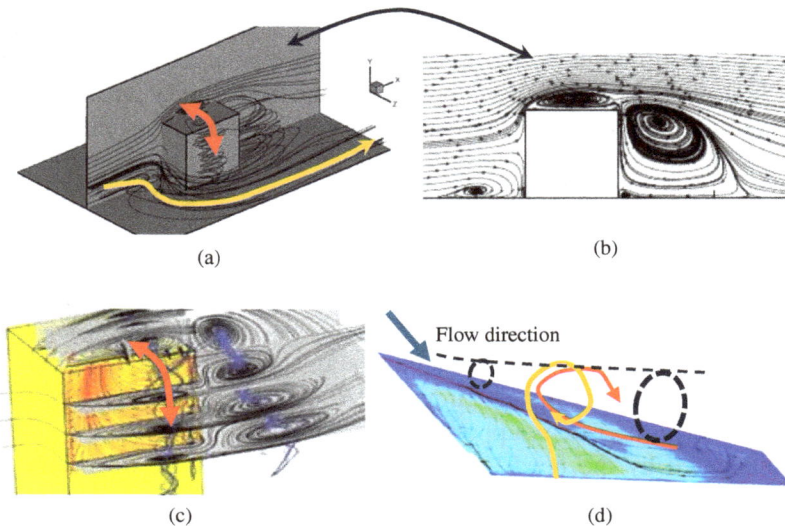

(a) (b)

(c) (d)

Figure 4.35. Some examples of three-dimensional separation: (a,b) flow around a cube on a wall; (c) flow around a tall square cylinder; (d) stall over the suction side of a swept wing.

Source: Yakhot *et al.* [18] (with permission); Breuer *et al.* [19] (with permission); Cao *et al.* [20] (with permission); Zhang *et al.* [21] (with permission).

surfaces, and the flow in this region interacts aerodynamically with the separations from the sides, as is schematically indicated by the red lines. In the case of the cube — indeed, any bluff body resting on a wall and approached by a near-wall stream — a recirculation zone is formed ahead of the cube because the high (stagnation!) pressure on the cube face forces the slow boundary layer over the bottom wall to decelerate and reverse, as explained in relation to Figure 4.20. This rotating vortex wraps itself around the base of the cube, forming what is called a "horseshoe vortex", roughly along the line indicated in orange in Figure 4.35(a). Therefore, the downstream wake is an amalgam of several vortices created on the sides, the roof and the bottom wall.

The third example, Figure 4.35(d), presents separation from a three-dimensional version of the geometry shown in Figure 4.25. On a real wing, there exists a highly influential spanwise flow component because the load on the wing and the pressure field vary substantially across the span. These distributions are especially affected by the sweep, the body of the aircraft, the wing tip and any winglet of the type shown in Figure 4.28. In essence, however, the image shown in Figure 4.35(d) indicates, in common with the two-dimensional case, the presence of stall due to the boundary layer in the front part of the wing being unable to withstand the adverse pressure gradient on the suction side, thus reversing its direction. The two obliquely colliding streams detach from the wing surface and form a three-dimensional recirculating vortex above the suction side. Here, it is important to point to the fact that this type of separation — a stall — differs radically, in terms of both structure and impact on the lift, from that shown in Figure 4.26 for the "delta wing", in which case the separation occurs at the sharp leading edge.

Separation around bodies subjected to boundary layers is an area of much interest in the context of city planning — in particular, when a new tall building is to be constructed within a densely built-up city centre. Any planning application for such buildings needs to be preceded by extensive wind-tunnel tests, often supported by computational simulations. These are mandated by the need to quantify the impact of the flow from any newly proposed building on surrounding buildings and also on the flow conditions close to the ground where pedestrians and cars need to move safely, especially in strong wind.

Two relevant images are shown in Figure 4.36. The left image shows a smoke test on a city-scale model in a wind tunnel. The flow is coming from the back and develops first along a bed of artificial roughness

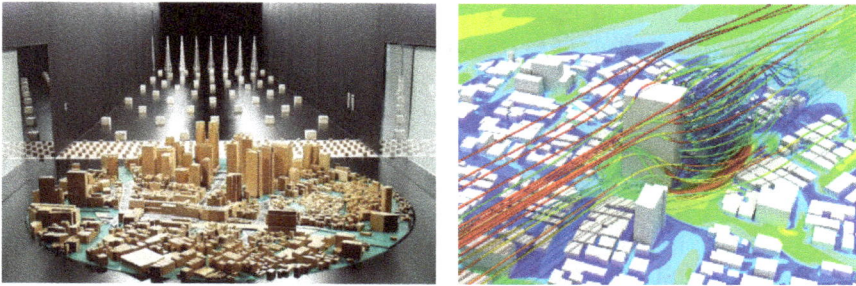

Figure 4.36. Wind-tunnel testing and computational simulation of building aerodynamics in the context of city-architecture planning.

Source: Shimizu Corporation (with permission); SimScale, www.simscale.com (with permission).

elements so as to create a realistic representation of an atmospheric boundary layer. Smoke tests of this type provide only qualitative statements, however, even if smoke probes are inserted into many key locations of interest, because smoke dissipates quickly in the highly three-dimensional, turbulent conditions that prevail in the complex geometry in Figure 4.36. Such visualisations are thus occasionally supplemented by advanced, but expensive, velocity measurements at key positions within the group of buildings. Computer simulations of the type shown in the right image can provide a far more detailed view of the flow between buildings, although great care has to be exercised, here too, to avoid over-rating the realism and accuracy of the results. For example, the computed flow will depend on the wind direction, the representation of the turbulence physics within the wind and the flow, and the spatial resolution of the calculation (refer to the discussion in Chapter 3). The most instructive feature in this image is the vortex created by the separation from the tall building in the rear of the image, and this process is clearly linked to the flows shown in Figure 4.35(a)–(c).

4.8 How Do Vortices Really Form?

This question might be an odd one to pose at this stage, following all the discussions and descriptions of separation, stall and recirculation in the previous sections. While we may claim that the discussion of many separated flows in this chapter has allowed us to acquire a degree of understanding of how recirculating vortices are formed, a valid counterclaim is that this understanding is rather conceptual and hazy. You may well ask the question: what process is responsible for the formation of the vortices in Figures 4.30 and 4.33? We may elaborate upon this question by also asking: knowing that a vortex is a rotating mass of fluid, where does this rotation within the mass come from and how are the vortices being sustained?

We know that a vortex is wrapped around a region of low pressure relative to that in the outer region. Recall that it is this pressure difference that is responsible for — or is consistent with — the curvature of the flow towards a low-pressure region. When we discussed the vortex in Figure 4.15 in Section 4.3, we asked questions about the pressure variation within the vortex, but we never asked why this mass rotates. This is an important question because the mass of rotating air must have "angular momentum", the same type of angular momentum that a rotating wheel has, and we know that to set the wheel in motion we need to apply a torque to the wheel, so there must also be a source that drives the rotation in the vortex.

To answer the question posed above, we need to first return to our simple boundary layer in Figure 4.19. It was pointed out then that the boundary layer is a sheared region manifested by the velocity gradient within it. However, this is not the only consequence of the shearing. Another important effect is that the shearing imparts "rotationality" onto the boundary layer, and this is crucially important to any vortex that is formed by fluid contained in a boundary layer.

To recognise the rotationality property — referred to as "vorticity" in the technical literature — we consider in Figure 4.37 the consequences of shear on an initially unsheared element of the boundary layer. By focusing on the diagonals of the element, we see that, as the element is sheared,

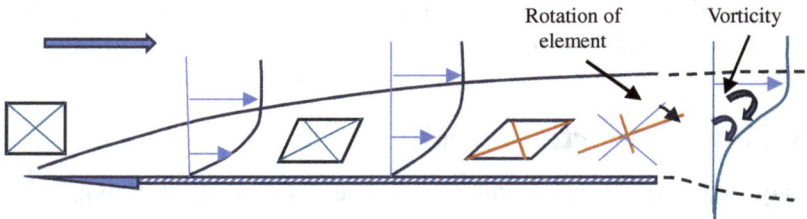

Figure 4.37. Illustration of rotation within a boundary layer growing along a stationary plate in an initially uniform stream. The flow separates from the plate, forming a free shear layer that contains the rotationality in the boundary layer.

it is also rotating. It is very important to appreciate that, once rotationality is imparted to the boundary layer, it normally remains within it. It can be shown that the rotationality in a boundary layer is changed when a pressure gradient is applied to it along the bounding wall. Hence, the shear imparted by a flat wall on a boundary layer in uniform pressure is the simplest mechanism by which vorticity enters a flow. However, we ignore pressure-induced vorticity variations in the following discussion and confine ourselves to the main phenomena.

What we conclude from the above simple argument is that our boundary layer is rotational. When this boundary layer separates, as it does in Figure 4.37, the rotationality is carried with the fluid into the separating mass, and this is the rationality that is embedded in any vortex formed with this mass. So, the main message to take away is that rotationality needs to be created at a solid surface (or liquid/gas interface) by shear — unless forced into the flow by a paddle or rotor — to impart rationality.

For a separating shear layer to give rise to a vortex, the presence of rotationality (vorticity) in the layer is not a sufficient condition. In fact, it is often emphasised in the technical literature that it is very important not to equate a vortex with vorticity; the two are linked but are different entities. If a boundary layer separates from the trailing edge of the flat plate without any constraints, disturbances, wavy instability or geometric confinement, the separated shear layer will progress and spread in the downstream direction with the vorticity embedded within it. To form a vortex, the shear layer has to be induced into a curved path, so as to wrap itself into a rotating mass, thus acquiring angular momentum. This can occur in various ways, dictated by the details of the flow conditions surrounding the shear layer or by geometric constraints. For example, in Figures 4.20 and 4.32, the lower-wall confinement gives rise to a low pressure within the corner region of the inclined and right-angled steps, respectively, and this pulls down the separated shear layer towards the wall, forcing part of it to form the vortex. In the examples of the unsteady vortex shedding shown in Figures 4.30 and 4.33, there is no physical confinement, but the lower and upper separated layers are alternately constraining each other, generating unsteady pressure fields that force both separating shear layers to curve and form vortices rotating in opposite directions, corresponding to the opposing orientations of the vorticity in the two layers. Moreover, as seen from the time-averaged flow fields for the round and square cylinders in Figure 4.34, the two vortices formed, on average, in the wake of the cylinders originate from the vorticity in boundary layers on the upper and lower sides of the cylinders, respectively, and the rear walls of the cylinders act in a similar way to the steps in Figures 4.20 and 4.32. Yet another way a vortex — and a powerful one — can form is by injecting a fluid tangentially into a circular chamber through circumferential slots. The purpose and benefits of doing so, in a technological context, will emerge from Section 6.5.

One further important fact to discuss relates to the orientation of the rotationality embedded in the fluid by shear. In Figure 4.37, the rotational property we call "vorticity" is in the plane of the paper. As noted, the vorticity remains in the fluid — it effectively becomes a property of the flow, similar to heat or a chemical species. Imagine now that we introduce a flow component normal the page and, additionally, make flow turn away from the normal direction, say upward or sideways or along any other curved path. It follows that the vorticity also re-orients with the curving flow. Hence, vorticity created in a boundary layer can be transported away from the boundary and finds itself embedded in a vortex that has an orientation different from the orientation of the vorticity-generating wall. A good example is the "horseshoe vortex" shown in Figure 4.35(a). The vorticity in this vortex originates from the boundary layer on the lower wall of the cube. This becomes embedded in the vortex formed at the upstream foot of the cube. The vortex is then bent into its horseshoe shape by the flow around the cube, and the vorticity, although influenced by pressure variations, essentially remains embedded in the vortex.

The above processes are pertinent to dust devils and fire tornadoes, for example, such as those shown in Figure 4.38(a,b). Also included in the figure is an image (c) of what most of us ordinarily associate with the term *tornado*, namely, the destructive twister formed within a "supercell" thunderstorm of the type most frequently observed in the "Tornado Alley" in the central region of the US. Importantly, however, this type of tornado is a much more complex phenomenon than a dust devil and a fire tornado, and it needs to be discussed separately, as is done later in this section.

In a dust devil and a fire tornado, buoyancy drives fluid upwards, although for different reasons — ground heating by the sun and combustion, respectively. In both cases, air is being drawn along the ground towards the location of rising air mass at which the pressure is low. This air will have acquired some angular momentum from the torque induced by the pressure minimum. In addition, fluid in the upper part of the tornado, which contains angular momentum, is moving radially outwards, then downwards towards the ground and is finally being ingested into the tornado along the boundary layer above the ground.

(a)

(b)

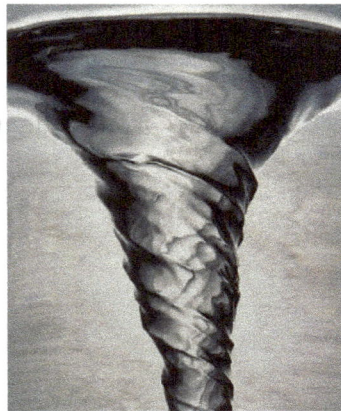

(c)

(d)

Figure 4.38. Rotating air masses: (a) dust devil; (b) fire tornado; (c) supercell thunderstorm tornado; (d) bath-tub vortex.

Source: M. Sekhomba, Wikimedia Commons CC-BY; J. Copelin, Instagram; J. Hobson, Wikimedia Commons CC-BY; R.D. Anderson, Wikipedia CC-BY.

The shearing motion relative to the ground adds vorticity to the air, and it is this vorticity, reoriented by the rising fluid, that becomes embedded in the core of the vortex in which the rotational speed is rising with radius, as is shown in Figure 4.15. As the outer air is being drawn along the ground into the tornado, it speeds up, racing faster and faster, due to the conservation of angular momentum. The core becomes surrounded by an outer layer in which the rotational speed is declining with radius and in which there is no, or very little, vorticity — a characteristic to which we shall return at the end of this section.

The thunderstorm tornado shown in Figure 4.38(c) arises from a complex — and not fully understood — clash between warm moist air that originates from a warm oceanic water body (e.g., the Gulf of Mexico) and cool dry air that originates from a cold land mass (e.g., northern Canada). This interaction spawns violent thunderstorms, intense rain and hail. As shown in Figure 4.39, a typical tornado-forming scenario includes three main elements: (i) a strong warm and moist updraft that is rotating due to the lifted air being spun by wind shear in the near-ground boundary layer, (ii) intense rain and hail in the contact area between the warm moist air and the cool dry air, (iii) and a separate vortex formed in a downdraft of a cooler air mass containing intense rain and hail. This vortex is driven by buoyancy-induced circulatory motions within the descending air mass. If this vortex is being entrained into the rising warm-air mass and pulled upwards, it is spinning at a faster and faster rate as it rises, ultimately manifesting itself as a tornado. This appears to start as a funnel within the upper part of the updraft and the cloud above it, which then descends towards the ground, eventually forming the twister shown in Figure 4.38(c).

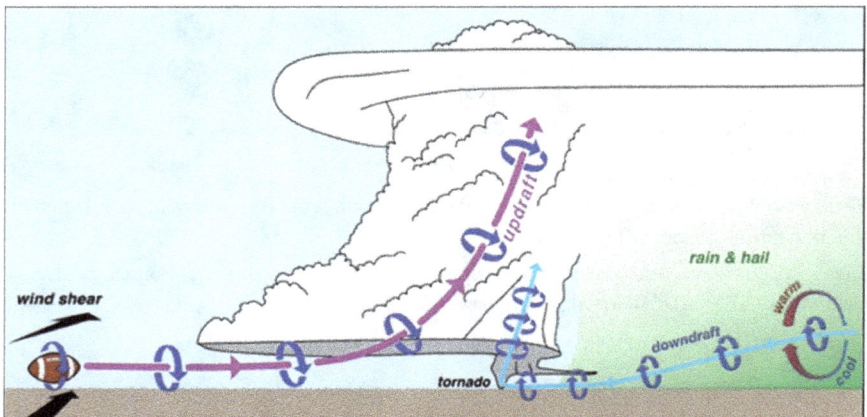

Figure 4.39. The processes involved in the formation of a supercell thunderstorm tornado.

Source: Markowski and Richardson [22] (with permission).

In the case of the bathtub vortex, shown in Figure 4.38(d), the vorticity can originate from the boundary layer on the bottom of the bathtub as the flow is being dragged towards the plug hole, or it can originate from, or be augmented by, vorticity and angular momentum imparted to the water during the filling process preceding the drainage.

While the hurricane shown in Figure 4.15 is also affected by the interactions described above — some pertaining to the supercell thunderstorm tornado excepted — this flow differs from those shown in Figure 4.38 (parts (a), (b) and (d)) in one important respect: a hurricane is a very large vortex and is strongly affected by the Coriolis force provoked by the Earth's rotation, see Figure 2.14, related explanations in Section 2.5 and the description relating to Figure 4.16 in Section 4.3. It is recalled, in particular, that the Coriolis force causes a body of fluid that flows horizontally over large distances to move along a curved path. In contrast, the small-scale flows in Figure 4.38 are compact and not more than marginally influenced by the Earth's rotation. Although the detailed aerodynamic as well as thermodynamic interactions giving rise to the formation and sustenance of hurricanes are complex, the vorticity generated in the sheared boundary layer along the ocean water surface is, here too, being reoriented and lifted upwards into the hurricane's eye by buoyant convection provoked by the warm ocean water. This vorticity finds itself embedded in the core of the hurricane within which the rotational speed decreases progressively from the destructive maximum to very low values.

We finally return to the radial profile in Figure 4.15, already mentioned in relation to Figure 4.38. Here, we consider a simplified model of the more complex vortex structure in the natural phenomena shown in Figure 4.38. The rotational velocity consists of two main parts: the inner part in which the fluid is assumed to rotate like a solid body — i.e., the rotational speed increases linearly with the radius — and the outer part in which the velocity declines in proportion to the radius. This structure is broadly representative of many vortices, especially in the natural world: the highest velocity is just outside the slow eye and then declines outwards. The rationality (vorticity) in the vortex is contained in the inner part, originating from interactions preceding the formation of the vortex — e.g., shearing in a boundary layer. The outer part is induced by the inner part and the pressure minimum in the centre of the vortex.

To understand the important difference between the two parts, consider the different behaviour of fluid elements in these parts, as shown in

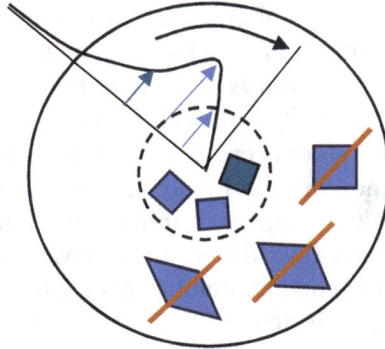

Figure 4.40.   Illustration of the behaviour of the fluid within the rotational inner core and the irrotational outer core in the vortex shown in the sketch in Figure 4.15.

Figure 4.40. In the core, the fluid elements rotate as if within a solid. In the outer part, the flow deforms, but the elements do not rotate — i.e., contain no vorticity. The existence of the outer region reflects the fact that the vorticity in the inner core only propagates slowly outwards, "nibbling" on the outer part by mixing. Without vorticity in the inner part, the fluid elements in the outer parts cannot rotate. Rather, if the action of the Earth's Coriolis force in large weather systems is set aside, outer fluid is being pulled into the vortex by the low pressure within it, and the shear exerted by the inner part on the irrotational outer fluid induces the outer flow which contains no vorticity.

4.9 Summary

At a very basic level, all that we need to know to be able to interpret the aerodynamic characteristics of any flow is the velocity — i.e., its magnitude (speed) as well its direction — and the pressure at any point in the flow and any instant in time. If we know both, we can deduce the streamlines, the mass-flow rate, the forces inside the flow and on the boundaries, and the momentum of the fluid across any area. This sounds easy but is hard in practice because both the velocity and the pressure vary greatly in space and (possibly) time. As will be seen later, in Chapter 5, matters are complicated substantially by the fact that the large majority of flows are turbulent in nature — i.e., consist of a complex mess of continuously evolving, interacting eddies that span a wide range of sizes.

If the flow is smooth — i.e., if we ignore the details of the turbulent motions — we can, at the simplest level, perform smoke visualisations in a wind tunnel or dye visualisations in a water flume to gain a qualitative view of the flow, in particular, the streamlines across sections in the flow. We have done so in a number of flows in this chapter, making observations and statements on where the velocity and pressure are high or low, and how both change as we move along streamlines. Furthermore, observing the curvature of the streamlines, we have inferred, again qualitatively, pressure trends normal to the streamlines. We could then identify separated and vortical recirculation zones, following an explanation of how a flow close to a wall — a boundary layer — slows down by the action of rising pressure, reverses its direction and then progresses to form a vortical region. We applied the insight we gained to the interpretation of flow behaviour around aerofoils, wings, car bodies, cylindrical obstacles and building-like objects, and we extended our consideration to the formation of different types of vortices.

If we want to go further and seek a more precise quantitative statement on the structure of whatever flow is of interest to us, we need to resort to elaborate and costly measuring systems and/or complex computer-prediction procedures, the latter solving the known equations that govern the velocity and pressure in flows. An indication of how this is done was given in Chapter 3. In this chapter, we have seen many images that were derived from such measurements and simulations, which helped clarify in detail what happened inside complex flows.

We have not discussed the equations governing fluid flows or computer codes solving the equations, for this would have defeated the whole rationale of this book. However, what we have done is introduce two fundamental concepts that also underpin the general flow-governing equations — namely, the conservation-of-mass principle and the force–momentum concept that links pressure forces to the velocity. These then facilitated our interpretation of features which we observe in flow images and of how the flows in question interact with their surroundings.

There are many practical flows in which we need to go beyond the basic fluid-mechanic description contained in this chapter. Specifically, in flows that include heat and species transport — e.g., in combustors — we also need to know the local temperature or the concentration of the relevant species (e.g., CO_2, N_2 and O_2) to determine how much heat or what amount of a particular species is being transported through any part of the flow or through the boundaries. We might be able to measure the temperature and concentrations, but this is, again, far from trivial. Here, too, more often than not, computer codes are being used to solve equations that govern the temperature and species transport. We shall return to the subject of heat transport in Chapter 6.

Chapter 5

Turbulence: Oh Dear!

5.1 Ignoring Turbulence Is Not an Option

Mention "Turbulence" and the image evoked in most people's minds is one of sitting in a plane that is being violently shaken by unexpected and, for some, dreaded atmospheric turbulence. We look at turbulence in flight in the following section. However, being tossed in a plane is just one of many manifestations of turbulence, a phenomenon which is extremely influential and wide-ranging in numerous fluid flows in nature and engineering.

Although we have repeatedly encountered turbulent flows in the foregoing chapters — just look back at some of the images in Chapters 1 and 2, including da Vinci's drawings in Figure 1.1 — we have, so far, judiciously and intentionally, avoided a discussion of the physics of turbulence and their consequences. Any expert in fluid mechanics would regard this as a remarkable feat because the large majority of flows (95%+) in engineering and the natural environment are turbulent. A flow has to be very slow, or very thin, or the fluid has to be very viscous to remain non-turbulent. We call this state "laminar".

There are not many opportunities to observe laminar flows. Pouring honey or Golden Syrup from a bottle results in a thin stream that is very smooth, lacking any ripples or roughness on its surface. This is one example of a laminar flow. Another, more difficult to produce, is a very slow water column discharged, say, from a kitchen tap with a long neck and smooth nozzle. This, too, demonstrates that a laminar flow is an exceptional state. Other laminar flows that cannot be readily observed are thin lubricating films and blood flows in thin vanes and capillary blood vessels. In contrast, the flow in major arteries — the aorta, in particular — and the flow in the heart are turbulent or close to this state. Nasal and lung flows are also turbulent.

To convey a quantitative measure that illustrates the rarity of laminar flow, we resort to the experimental observation that a pipe flow transitions from a laminar to a turbulent state when the group (*velocity* × *diameter* × *density/viscosity*) reaches the value of around 2300. If we assume that the fluid in the pipe is water and that the flow velocity is only 10 centimetres per second, the pipe is required to have a diameter smaller than 23 millimetres for the flow to remain laminar. The large majority of engineering flow and virtually all environmental flows are much more vigorous and/ or have much larger dimensions and are therefore turbulent. Even very slow flows, which we might expect to remain laminar, may become turbulent if disturbed by sharp geometric features that cause separation and recirculation.

It is instructive to pre-empt the multifacetted cause-and-effect arguments presented in this chapter by pointing to one key consequence of turbulence — namely, loss of useful energy. This is due to the chaotic eddying motions causing intense shear and friction between and within the turbulent eddies, which then leads to some of the pressure and kinetic energy within the flow being transferred to useless heat. In this singular respect, turbulence is a damaging process, and its suppression has been the objective of sustained efforts based on a variety of creative flow-control techniques, a subject discussed in Chapter 7. On the other hand, turbulence has the advantageous attribute of being an effective mixer of flow properties, such as heat, momentum and concentration of chemical species, particles and droplets. Hence, turbulence presents us with a problematic, often delicate, balance between good and bad that depends greatly on the purpose or function of the flow in question.

5.2 Tumbling and Toppling: Turbulence in Flight

Even more than "stall", atmospheric turbulence, illustrated superficially in Figure 5.1, strikes fear in any nervous flier. Yet, aircraft crashes due to turbulence are extraordinarily rare, far rarer than due to stall or pilot error, because commercial aircraft are designed to withstand extreme atmospheric conditions.

Figure 5.1. Atmospheric "clear-air" turbulence generated at the interface between fast- and slow-moving streams of air, which can also flow at any angle relative to each other. The aircraft experiences turbulence as it moves through the interface layer.

Source: © PilotMall.com Inc (with permission).

In a recent incidence, on May 22nd 2024, a Singapore Airlines plane suddenly dropped 2000 metres within a couple of minutes after encountering clear-air turbulence near the Equator. Although some 70 passengers were injured, a few severely, and much internal mess was caused by the drop, the injuries were due to passengers not wearing their seat belts and being flung towards the ceiling or being hit by flying objects. Importantly, the aircraft itself sustained no structural damage and landed safely in Thailand.

If an aircraft manages to avoid clear-air turbulence, the calm of a nervous flyer may be shredded by a bumpy landing phase when the aircraft is flying through clouds or through the "atmospheric boundary layer". The former is a consequence of thermal updrafts, discussed in Section 5.3, while the latter is caused by wind in the lower part of the atmosphere creating a sheared boundary layer of the form indicated by the left-most profile in Figure 4.17. This is, fundamentally, no different from the shear profile shown in Figure 4.37, and the turbulence-generating mechanisms are the same, as described below.

Turbulence is also occasionally experienced while the aircraft flies over a high mountain range, e.g., the Alps. This is due to high winds across the mountain range provoking flow separation from the sides and tops of the mountains, which then gives rise to eddies and vortices of the type shown in Figures 4.8, 4.17, 4.30, 4.33 and 4.35. Wave-like and pressure disturbances induced in the atmosphere by these turbulent motions

Figure 5.2. "Wake Turbulence" comprising well-defined vortices that are generated at the tip of lifting wings or slats and persist for long distances behind the aircraft.

Source: Shutterstock (with permission); Pilot Institute (with permission).

can propagate to substantial heights, but these are rarely severe or of any consequence to the safety of the aircraft.

Yet another type of turbulence, familiar mainly to aircraft buffs, is "wake turbulence", shown in Figure 5.2. The term is put between quotes because the phenomenon does not really comply with the technical definition of turbulence, as will become clear below. Rather, wake turbulence signifies the shedding of vortices from the tips of wings or flaps. These arise from a circulation around the wings or flaps, implied by the lift they generate, which is shed from their tips in the form of vortices. A simpler, though imprecise, explanation is that the vortices are generated by air moving from the (lower) high-pressure side of the wing or slat to the (upper) lower-pressure side across the tip, as indicated in the right-hand part of Figure 5.2. These vortices will eventually break down into a chaotic set of eddies with the vortex becoming increasingly diffuse, but it is the well-defined vortex pair shown in Figure 5.2 that is popularly known as "wake turbulence".

5.3 What Makes Turbulence Tick?

First, before we consider why turbulence is important to us, apart from the loss of useful energy and the mixing attribute singled out already, we have to ask the question: what actually is turbulence? Surprisingly, this is a very hard question to answer, despite 150 years of scientific research. We can observe and even predict its onset, evolution, properties and effects, but we have a weak understanding of the fundamental mechanisms that are at play. In fact, turbulence is said to be the last unsolved problem in classical physics, and some great minds — including Albert Einstein — shied away from trying to unravel its origin. Werner von Heisenberg, the pioneer of quantum mechanics, is reputed to have said:

When I meet God, I would ask two questions: why quantum mechanics and why turbulence? I believe God would be able to answer the former.

So, what is turbulence? The scientist's best reply might be: "it's an insta-bility that results from the nonlinear character of the equations that govern fluid flow". This will not satisfy the large majority of the readers of this text. Therefore, rather than pursuing the subject from this perspective, we shall focus on what we can observe.

Turbulence is generated by shear (and other drivers, covered later). In the simplest of circumstances, shown in Figure 5.3, two layers of fluid move relative to each other at different speeds — i.e., the interface is being sheared. Unless the speed is very low and the viscosity is very high (honey or highly viscous oil moving very slowly), the interface begins to deform into a wavy pattern. These waves grow, begin to fold and develop into vortices, the attendant vortical curvature being promoted by pressure peaks and troughs that are caused by the instability.

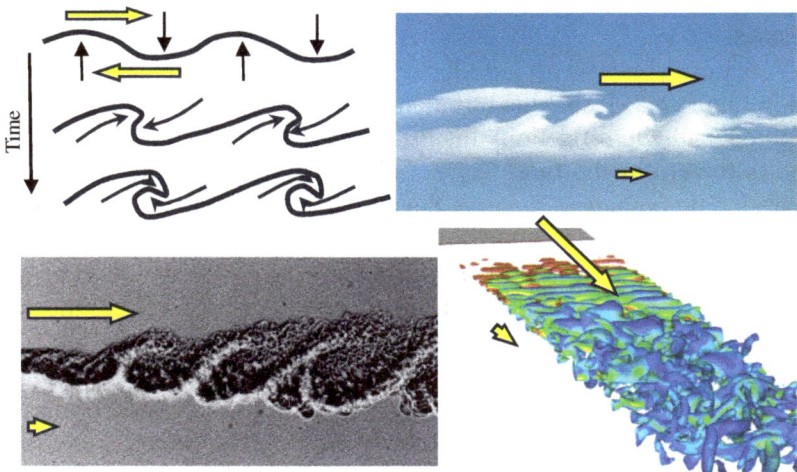

Figure 5.3. Instability in a free shear layer that results in large, ordered vortices, followed by breakdown into chaotic turbulence.

Source: B. Martner, NOAA/ETL CC-BY; Van Dyke, *Album of Fluid Motion*; Kok and v.d. Ven [23] (with permission).

With luck, we can observe the process in the atmosphere when two layers of air move relative to each other with the lower layer containing wispy clouds (Figure 5.3). If the shear persists and is sufficiently strong, these vortices become increasingly irregular, also in the spanwise direction — normal to the page — and begin to break into three-dimensional smaller vortices. "Break" is imprecise: large vortices will generally persist, continuing to be generated by the shear, but will co-exist with an increasing array of smaller vortices, the result being a chaotic unsteady field of continuously evolving, three-dimensional, interacting vortices of many sizes. The above interactions also apply, in principle, to the clear-air turbulence shown in Figure 5.1, although in that case, turbulence may be very intense and feature large powerful eddies.

A slightly more complicated scenario arises when a uniform non-turbulent ("laminar") flow approaches a flat plate, as shown in Figure 4.19. The flow is then subject to shear close to the surface. For a while, the flow is ordered and remains stable. As the shear layer thickens, wavy instabilities set in, as indicated in Figure 5.4. Initially, these are two-dimensional — i.e., they do not vary in the spanwise direction — but they eventually become three-dimensional and begin to develop "hairpin vortices" in spots and increasingly large patches close to the wall. As the flow progresses further, the regular pattern breaks down, becoming increasingly chaotic, and the result is, again, a chaotic field of vortices of different sizes.

A feature of near-wall turbulence, relative to the unconstrained condition in Figure 5.3, is that the turbulence intensity tends to drop as the wall is approached. On reflection, this is not far-fetched: at the wall, the speed reduces to zero as the fluid has to "stick" to the wall (refer to an earlier discussion on boundary layers in Section 4.4). This zero value applies not only to the overall (average) speed of the flow but also to all speed fluctuations. Hence, at the wall itself, turbulence vanishes. At the same time, the wall has a blocking effect on fluctuations normal to the wall. Specifically, pressure fluctuations associated with wall-impinging turbulent motions are reflected by the wall towards the eddies in the vicinity of the wall, and this causes the damping of wall-normal fluctuations. The two processes thus combine to result in a progressive weakening of turbulence as the wall is approached. Flyers of a nervous disposition, who follow keenly the intensity of turbulence as the aircraft approaches the runway, will be aware of the relative calm — unless the weather is stormy — before the landing gear touches the ground.

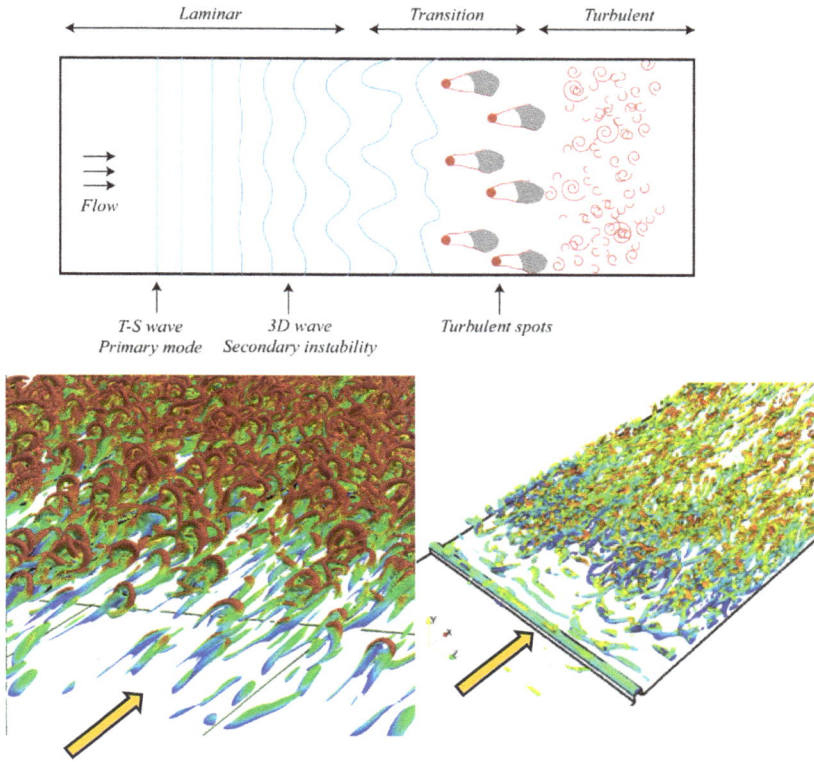

Figure 5.4. Instability in a boundary layer on a flat plate that results in waves, followed by spotty transition to a "forest of hairpin vortices" and eventually breakdown into chaotic turbulence.

Source: Veerasamy [24] (with permission); P. Schlatter and R. Örlü (private communication); Chen and He [25] (with permission).

The process of progressive breakdown of vortices in any turbulent flow is elegantly described by Lewis Fry Richardson, a pioneering theoretical meteorologist, in 1922:

Big whirls have little whirls,
That feed on their velocity;
And little whirls have lesser whirls,
And so on to viscosity.

The first part says that the big eddies are generated by the shearing motion — "*....feed on their velocity*". The second part is more enigmatic.

Here, we need to appreciate first that the breakdown of bigger eddies into smaller ones means that energy has to be taken away from the big eddies to feed the smaller eddies. In other words, there is a kind of "cascade" of eddy sizes and energy. As the eddies become smaller and smaller, they are increasingly damped by the action of viscosity — say, very small eddies cannot exist in highly viscous oil. Thus, this is the meaning of the second part: the energy cascades down to the smallest eddies, which are then killed off by friction with their energy dissipated into heat (although, the temperature rise involved is generally too low for us to sense the dissipation process).

This is as far as we shall take the question of "what is turbulence?", and we turn our attention to the second question; "why is it important to us (flying aside)?" In the process of addressing this question, we shall also argue that the material presented in Chapters 3 and 4 regarding shear flows, separation and recirculation is not invalided by the presence of turbulence but merely needs to be qualified and re-interpreted.

It has been pointed out that an important effect of fluid viscosity is to cause the shear layers to spread, for example, the boundary layer in Figure 4.19. This spread reflects the mixing of momentum by the action of molecular (Brownian) motion. This same mixing property is also responsible for mixing heat — i.e., manifested by the spread of the temperature — and also mixing and spreading of species concentrations — e.g., smoke and CO_2. However, because the viscosity is low in most fluids, this process of mixing is weak, and in air and water, it is very weak. The very long and thin contrails left behind high-flying jets in a quiescent, non-turbulent atmosphere, as shown in Figure 5.5, illustrate the slow spread by viscous effects.

Turbulence is a hugely important contributor to mixing. This was first implied in this book by Figure 2.4, showing the flow as we breathe out air from our lungs. To make this point in simple terms, we consider two adjacent layers of two different gases — say N_2 and O_2 — in Figure 5.6.

By introducing turbulent fluctuations across the interface, we cause two packets of fluid of the same mass to be exchanged between the two gases. Imposing this exchange for many other packets with a variety of fluctuations normal to the interface is what we mean by "mixing" — look again at Figure 2.4. The process of turbulent mixing increases hugely the contact surface between the two gases, allowing viscous mixing to occur across the convoluted boundary of packets of dissimilar composition in contact.

Figure 5.5. Long, thin, slowly dissipating contrails left behind by jet aircraft flying in quiescent atmospheric conditions.

Source: NOAA National Weather Service CC-BY.

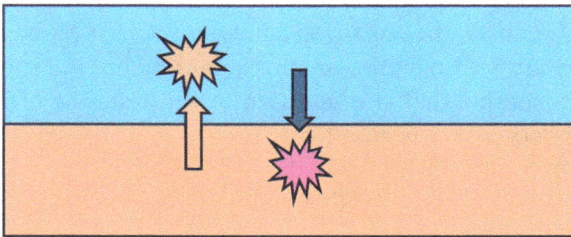

Figure 5.6. A simple model illustrating the mixing by turbulent fluctuations, caused by mass and property exchange across the interface of two different fluids.

5.4 The Good and Bad of Turbulence

We can now return to our discussion of shear layers, vortices, aerofoils, separation and recirculation to ask the question: are all the descriptions valid in view of the complication introduced by turbulence? The answer is a qualified "yes". The qualification rests on the concept of looking at turbulence from a statistical perspective. What does this mean? It means that we turn our attention away from the unsteady, multi-scale nature of the turbulence field and refocus on the effects of turbulence on the time-averaged flow. We argued in favour of doing so earlier when we related the average flows in Figure 4.34 to the unsteady features in Figures 4.30 and 4.33. If this averaging point of view is accepted, one essential effect of the enhanced mixing by turbulence is that shear layers spread at a higher rate than purely by viscous action (Brownian motion). Another effect is that a boundary layer approaching a region of high pressure is less sensitive to this increase and remains attached for longer. Hence, in the case of separation and stall, we observe that both are delayed by the enhanced turbulent mixing of momentum within the boundary layer. Yet another feature is that recirculation zones are shorter because the separated shear layers reattach earlier due to their ability to recover faster.

Why is a turbulent boundary better able to resist the effects of rising pressure? To answer this question, we return to Figures 4.20 and 4.24 in Section 4.4. There, we argued that separation occurs because the near-wall flow is slow, has little momentum and is unable to resist the rising pressure for long. Hence, it follows that any process that increases the momentum of the near-wall layer will inhibit separation. This is precisely what turbulence is doing: it mixes momentum. It transports high-momentum fluid packets from regions remote from the wall towards the wall, by wall-normal fluctuations, thus increasing the momentum of the fluid close to the wall. Analogously, the enhanced spreading rate of a free shear is also due to momentum redistribution by mixing normal to the shear layer.

A classic example that is often used to illustrate the effect of turbulence on boundary layers is a golf ball, as shown in Figure 5.7. When the ball is smooth, the boundary layer developing on the front of the ball is not turbulent because it is very thin and has not yet been affected by the wavy instabilities shown in Figure 5.4. The boundary layer is unable to resist the area of rising pressure behind the ball as the flow begins to decelerate. When the ball surface is dimpled, however, turbulence is triggered in the boundary layer, and the flow remains attached over a portion of the rear surface of the ball. What is the benefit? To answer this question, we return to Figure 4.8 in Section 4.2 to identify an important consequence of the flow becoming turbulent. When discussing this figure, it was pointed out that a perfectly symmetric flow around the vertical centreline of the ball implies zero drag. While this is not possible in reality, any process that decreases the level of asymmetry reduces the drag. Hence, counterintuitively, the dimpled ball will fly further than the

(a)

(b)

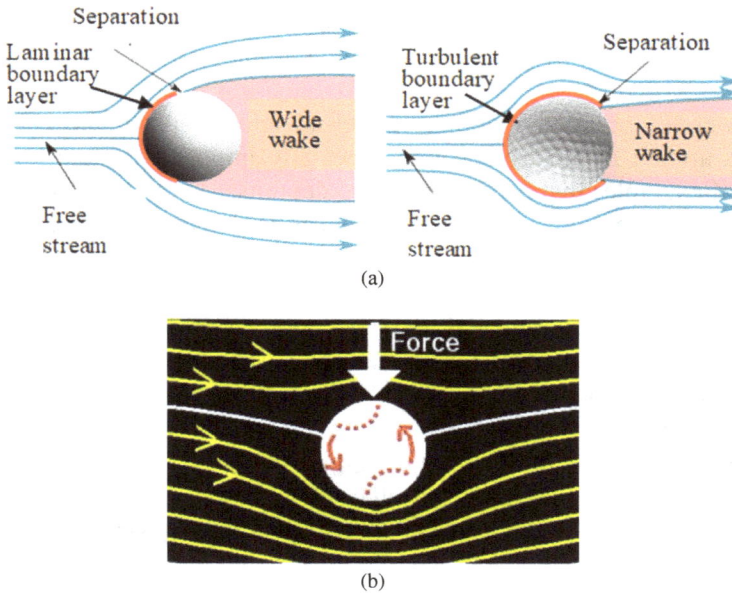

Figure 5.7. The flow around (a) a smooth and pitted golf ball and (b) a rotating tennis ball.

Source: Reddit; NASA.

smooth ball. The level of friction drag on the surface of the pitted ball may be higher because of turbulence, but the drag reduction due to the reduced front-to-rear pressure difference more than compensates for the increase in friction drag. The same applies to a football or cricket ball.

As an aside, it is remarked here that the curved path followed by a football, if cleverly hit (familiar to those who have watched the film *Bend It Like Beckham*), or a tennis ball, is not connected to turbulence, but is a consequence of rotation imparted to the ball, as shown in Figure 5.7(b). This rotation creates an asymmetry in the flow, which then results a force normal to the main direction of travel — referred to as the "Magnus Effect". The curved flight of a cricket ball involves more complex interactions. The spin of the ball imparted by the bowler is one of several factors. Another is the orientation of the rough seam on the periphery of the ball relative to the direction of flight. A third factor is that roughening one side of the ball and polishing the other favours a sideways component to the flight of the ball, and this is due to the flow along the rough side being more turbulent (or more extensively sheared) than on the smooth side, thus causing flow asymmetry relative to the vertical centre plane of the ball.

From the point of view and heat and mass transfer, the enhanced mixing arising from turbulence is especially important. There are hundreds of examples that can be fielded to demonstrate this. Figures 2.10 to 2.13, for example, illustrate that turbulence is crucially important to the dilution of high-concentration discharges and toxic substances. This is not only so in atmospheric flow but also in rivers, lakes and oceans in which turbulence is also instrumental in oxygenation and redistribution of nutrients. Turbulence in rooms enhances the distribution of heat and the exchange of oxygen and CO_2. Near-wall turbulence in pipes is crucially important to the effectiveness of heat exchangers, boilers and heat pumps. Combustion relies on turbulent mixing, as seen in Figure 2.11, unless it is of very modest yield, as is the case in a candle flame. An amusing "advantage" of turbulence is that insects dislike flying in turbulence, not because they are afraid of it but because of the high unsteady accelerations to which they are subjected within the small but intense vortices.

Some important disadvantages of turbulence should not remain unmentioned. The transfer of useful energy to useless heat has already been highlighted. Turbulence at solid surfaces substantially increases the friction drag. The reason is indicated in Figure 5.8.

Following the initial non-turbulent stretch, the flow transitions to a turbulent state, and the boundary layer spreads at a higher rate due to enhanced mixing of momentum. More to the point, the turbulence in the boundary layer away from the immediate vicinity of the wall flattens the velocity profile in the boundary layer — again, a consequence of enhanced momentum mixing. The velocity has to decline to zero at the wall, and the shear gradient at the wall increases significantly, which increases the frictional drag. This is especially troublesome in the transport of water and oil

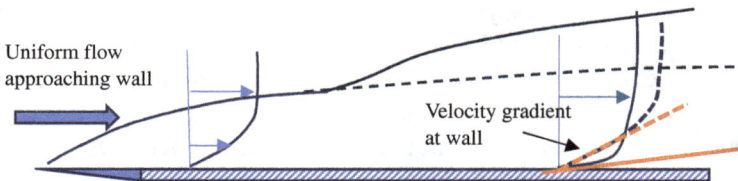

Figure 5.8. Differences between a non-turbulent (laminar) and a turbulent boundary layer (note the strong increase in shear at the wall in turbulent conditions).

in long pipelines, as this requires a major expenditure in power to pump the fluids against the frictional resistance.

The increase in the loss of useful energy due to turbulence decreases the efficiency of engineering devices, such as compressors and turbines. The reason is that, to sustain the turbulent eddies, energy has to be extracted from the flow, and this causes a decrease in the amount of energy in the flow that is available to do useful work. As explained earlier, this energy then cascades down the eddy-size range to eventually dissipate as (useless) heat by viscous action.

Let us finally return to the fear of turbulence in the context of flying. We have already discussed some aspects of two types of turbulence: clear-air turbulence at moderate-to-high altitudes and turbulence close to the ground during landing. The third type of passenger experience occurs when the aircraft passes through clouds, especially thick cumulus clouds. The first is the most insidious, as it tends to occur suddenly and without warning in cruise conditions when many of us release our seat belts. The fact that the shear rate is high and the layer is unconstrained favours large eddies and strong vertical turbulent fluctuation. The aircraft can thus drop or/and rise by a distance of dozens of metres in a very short amount of time. The instructions to passengers to keep their seat belts fastened at all times is, therefore, not an empty recommendation. Turbulence during landing is rarely severe because of the relatively low wind speeds involved and the proximity of the ground, for reasons explained already. There are some special circumstances in which turbulence during landing can be risky, first and foremost in strong thunderstorms when turbulence is augmented, or even dominated, by strong updrafts and downdrafts generated by storm cells, in which case shear-induced turbulence may be subordinate.

The third type of turbulence, mentioned only in passing in Section 5.1, is driven by buoyant updrafts and downdrafts, plumes and thermals, as shown in Figure 5.9. These arise from thermal heating near the ground with warm packets of air — plumes and thermals — rising upwards and colder packets sinking. As the warm packets rise, the moisture condenses and forms clouds, a process that releases heat and makes the packets become even more buoyant, thus rising even faster. These vertical motions are large, distinctive and can be fairly organised, so they do not quite qualify for being designated as "turbulence", which is strictly a chaotic multi-scale state. Rather, turbulence is created by intense shear between updrafts and downdrafts, and between the up/

Figure 5.9. Clouds, thermal updrafts and downdrafts interpreted as turbulence during flying through clouds.

Source: Shutterstock (with permission); Lülff [26] (with permission); E.P. v.d. Poel and R.O. Monico https://visit-dav.github.io/visit-website/examples/ex42/.

downdrafts and wind shear. However, the vertical motion of the aircraft, felt as it crosses the clouds, is mostly a reaction to the vertical buoyant motion of the packets in the plumes and thermals. In adverse, unstable, weather conditions, updrafts and downdrafts can become severe, being accompanied by thunderstorms, intense precipitation and strong shear. This can readily occur around the tropics, but even then, the risks to aircraft safety are low.

The process shown in Figure 5.9 may be described as buoyancy-induced turbulent mixing, wherein low-density fluid packets at the bottom rise while high-density fluid packets descend. This is an unstable environment — referred to as "unstable stratification" — because the density field provokes and amplifies fluctuations. There are several natural phenomena that arise when the density field is reversed — i.e., when the density in the lower layer is high and that in the upper layer is low — a condition referred to as "stable stratification". In such a condition, turbulent fluctuations and mixing are damped, possibly to the extent of vanishing. The most frequently observed manifestation of this process is the formation and persistence of near-ground fog in winter due to the near-ground temperature being low and the temperature in the layer above it being higher. This fog can be extremely reluctant to dissipate unless the ground is progressively heated by sunshine, in which case a local density reversal very close to the ground causes slow turbulent mixing, referred to as "penetrative convection".

Another important example is the absence of mixing in deep water bodies (e.g., the Black Sea) due to the deep water being cold and the top layer being warm. Mixing ceases in a thin layer separating the upper warm and lower cold water, referred to as a "thermocline". The importance of this stable stratification lies in a serious depletion of oxygen in the lower layer, with aquatic life unable to be sustained below the thermocline.

A slightly more complex set of interactions, to which the process just discussed contributes, arises when a warm plume rises in relatively quiescent atmospheric conditions in which the ambient temperature increases with height, i.e., in stable stratification. Such a condition is shown schematically in Figure 5.10. We assume here that the material properties of the plume are the same as those of the surrounding air. As the plume rises, it mixes with atmospheric air and its temperature drops. At the same time, the atmospheric temperature is rising. This continues until the plume temperature is the same as the ambient temperature. At this point, buoyancy

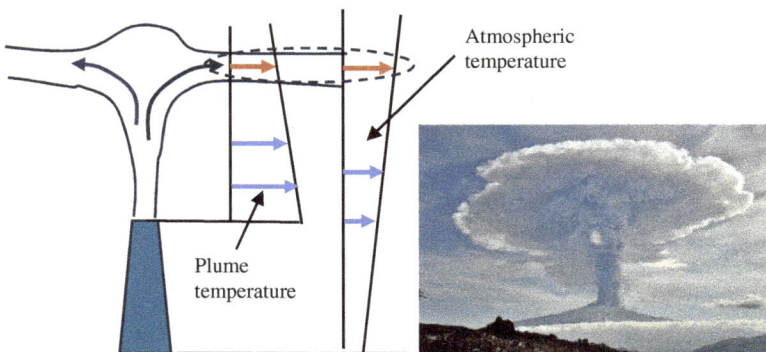

Figure 5.10. Warm plumes rising in a stably stratified atmospheric conditions: (a) schematic used to explain the mechanisms; (b) a volcanic plume discharged by Mt. Etna. *Source*: de' Michieli Vitturi and Pardini [27], CC-BY.

is lost, the plume stops rising, and it spreads sideways. Due to the stable stratification in the atmosphere, turbulence in the spreading plume is gradually suppressed, and the plume spreads as a smooth cloud. The dome above the centre of the spreading cloud is a consequence of the upward momentum of the plume prior to its reaching the thermal equilibrium height. The plume fluid in this region continues rising for a short distance in a negative buoyancy condition in which the buoyance force is directed downwards against the momentum of the plume. Figure 5.10 also contains an image of a volcanic eruption to which the mechanisms just described also apply.

To understand the mechanism by which stable stratification damps turbulence, we return to Figure 5.3. Imagine that the lower layer has a higher density than the upper layer — a case of stable stratification. Next, consider an upward-directed fluctuation that transports the high-density packet into the lower-density layer. Buoyancy produces a downward force on the packet, thus pushing it down again. An entirely analogous interaction occurs in respect of the lower-density packet in the upper layer finding itself in the higher-density lower layer. Thus, we conclude that stable stratification damps turbulence, and if the stable stratification is sufficiently strong, turbulence is killed off altogether.

5.5 Summary

Turbulence is a uniquely intriguing and influential phenomenon in fluid flow. It is characterised by a tangled forest of eddies of many scales, of size as small as a fraction of a millimetre to a size not much smaller than that of the flow as a whole — tens of metres in the case of fast and sheared atmospheric flows. Turbulence causes the flow to change rapidly in space and time in a manner that never repeats itself and can therefore not be described in closed form. Neither is the unstable evolution of turbulence from a smooth-flow state to a fully turbulent state properly understood, despite many years of study. What we can do is to observe turbulence as it evolves — which is rather different from understanding it — monitor large fluctuations, if these are of interest, and examine its influence on natural phenomena and engineering applications. We do so, in most circumstances, by focusing on the time-averaged behaviour of the turbulence-containing flow rather than attempting to describe the time evolution of the turbulence in detail. A time-averaged point of view is equivalent to taking a long-time exposure of the flow or taking many snapshots of the flow and superposing them to get an average.

Turbulence gives rise to a mix of life-critical advantages and a few penalties. One crucially important advantage is that it is an excellent mixer: it acts to disperse smoke, toxic substances, vapour, fog and heat; it is essential to combustion, as it mixes fuel with oxygen, preventing extinction by the accumulation of burnt reaction products and allowing the flame to be sustained; it also redistributes momentum, reducing velocity variations across sheared regions. This last property is very influential to the onset of separation from surfaces subjected to deceleration due to pressure-induced resistance against the flow, the prevention of stall on aircraft wings being one important example. Turbulence is also extremely important for cooling hot surfaces, evaporating water at liquid-air interfaces and for distributing moisture, CO_2 and oxygen in the atmosphere and oceans. Hence, however challenging turbulence may be to our understanding of the behaviour of fluid flows, without turbulence, life as we know it would be impossible.

Chapter 6

The World Is Full of Life-Sustaining Flows

In the course of discussing many important flows and flow phenomena in previous chapters, we have already encountered numerous practical scenarios in which these phenomena play a key role, including aircraft, cars, buildings, chimneys and rockets. Many natural flow phenomena have also been discussed, including weather systems, volcanic eruptions, bird flight and physiological flows. Yet, these constitute but a tiny fraction of the enormous number of environmental systems and man-made machines and processes which govern our lives and on which we depend daily. In fact, there are very few situations in our everyday life that do not involve fluid mechanics — from stirring sugar and milk in a cup of coffee to mixing concrete and water in construction, treating sewage, regulating rivers, melting metals, manufacturing drugs, keeping patients alive with heart/lung machines and creating complex shapes by fluid-jet printing. It would be entirely impossible to cover even a major proportion of the flows that underpin natural phenomena or are exploited in engineering applications. All that can be attempted in this chapter is to add, modestly, to the configurations already discussed and give added exposure to the ubiquity and complexity of fluid-flow-based, man-made exploitations of fluid flows.

Numerous important engineering applications rely on the extraction of power from moving fluids or on imparting power to propel fluids. Many of these involve aerofoils, blades and wing-like components that function, in principle, in the manner explained in previous chapters: they create lift and produce a gain in useful mechanical energy — good, but they also cause energy loss by friction due to the relative motion between the fluid and the streamlined body — bad. The lift force is used to drive a wheel or an axle that links to an electrical generator or a mechanical system or is used to propel fluids by applying torque to a wheel that contains aerofoil-like blades. Gas turbines are used to power aircraft by jet thrust, aircraft and ship propellers, helicopter rotors and electric generators. The auxiliary gas turbine installed at the back of civilian jet airliners and operating while the aircraft are parked on the tarmac generates electrical power that provides lighting, ventilation and air-conditioning to the plane. Propellers, fans and rotors consist, themselves, of collections of wings or blades. Rotary pumps and compressors use blades to increase the pressure of gases and to drive water to higher elevations in reservoirs, or to pump fluids through pipe systems, in domestic water-supply networks, in oil refineries, through nuclear-reactor cores and in sewage-treatment plants. Water turbines convert the potential energy stored in the "head" (elevation) of water behind a dam into mechanical and then electrical energy. Tidal turbines similarly convert the energy in a stream of sea water into electrical energy. Wind turbines generate electricity by wind that causes a lift force on the turbine blades and hence torque on the blade-carrying shaft, which then rotates to drive an electrical generator. Cooling fans are used in numerous applications to cool equipment, including millions of computers, ventilate rooms and also heat them when incorporated into fan heaters and heat pumps.

The long list given above merely conveys an inkling of the virtually endless number of technological applications and systems that exploit the properties of fluids — and this only in the rotary environment. The predominance of wing-like components in many fluid-engineering machines is the reason for the inclusion of several rotating fluid-flow applications in this chapter. The idea is not to give a comprehensive overview but to convey an appreciation for what fluids do as they interact with rotating arrays of blades in the context of power management. In addition, this chapter includes descriptions of heat-transfer applications, medical applications, metal casting, cyclone-flow devices, water-distribution networks, wave-motion on water surfaces and two related recreational activities: one that exploits an intriguing phenomenon in tidal rivers akin to a breaking wave and another that hinges critically on wind flow, namely, sailing.

A significant number of the flows covered in this chapter deal with power generation and energy management and thus have clear implications for the environmental challenges highlighted in Chapter 1. Several flows also involve aspects of flow control and manipulation, designed to achieve a particular behaviour or characteristics advantageous to their operational performance. However, these will not be identified explicitly as such in this chapter. Rather, Chapter 7 will pick out the specific facets that are relevant to control and discusses them in the context of a range of flow-control techniques.

6.1 Power to My Jet: Gas Engines

Although civil aviation accounts for less than 3% of global CO_2 emissions, relative to around 30% from all sources associated with transportation, this sector has acquired an arguably unfair level of notoriety as a major contributor to global warming. It turns out that there is hardly any other industrial sector that has done more to reduce the harmful effects arising from carbon-based fuel consumption. This is conveyed strikingly in Figure 6.1, which shows fuel-consumption data per unit thrust for civil aircraft entering service over the past 65 years. The graph contains two sets of data points: the top set of red dots showing fuel consumption is raw form, as a percentage change relative to 100% for the Comet 4 aircraft, while the lower set of black dots indicates the reduction in terms of fuel consumption per passenger-mile. In effect, the latter set is an integral indicator of the advantage arising from all efforts to improve the efficiency of civil airliners. To be absolutely clear, transporting a passenger in 2015 on a modern Boeing or Airbus plane requires only 25% of the fuel per mile needed in 1960.

Much of the phenomenal reduction shown in Figure 6.1 is the result of improvements to the aerodynamics, materials and combustion systems in jet-engine technology. However, efforts in many other areas have made significant contributions, too, including the aerodynamics and materials of aircraft bodies and especially of wings (see Figure 4.28 and the related discussion). All these improvements were only possible because of research into, and analysis of, fluid-flow fundamentals and of the interaction of flows with the solids in contact with the flows. Two major contributors are shape optimisation and flow control, targeting drag reduction, and the increased use of composite materials in wings. We shall return to the former topic in Chapter 7.

The operational principle of a gas (jet) engine is simple: compress air in a rotating compressor, direct it to a combustor, add fuel, burn the

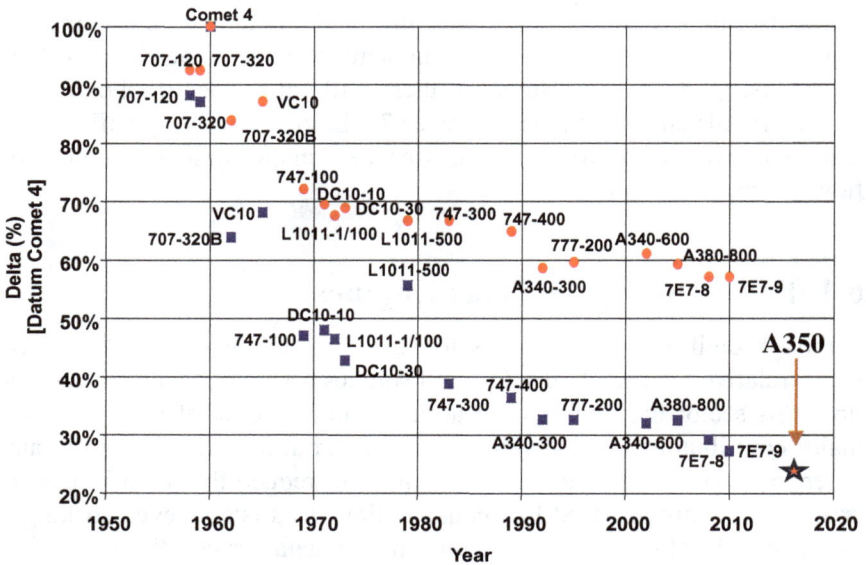

Figure 6.1. Reduction in fuel consumption per unit engine thrust of civil airliners relative to the Comet 4 airliner in 1960. The upper set of data points gives the reduction in raw percentage terms, while the lower set gives the reduction in terms of consumption per passenger-mile.

Source: Adapted from IPCC [28]; last data point added by author.

mixture, expand the gas in a turbine and use part of the power to drive the compressor and the rest to produce a jet or drive another device — a rotor, a fan, a propeller or an electrical generator.

The jet engine shown in Figure 6.2 is one of many used for large civil airliners. Characteristically, the engine contains a large fan that creates a moderately fast but high-volume-rate air stream in a bypass duct, which discharges from the back of the engine together with the noisy hot-gas jet from the engine core. The non-experts might assume that the thrust is produced mainly by the hot jet issuing from the turbine-exit nozzle. In fact, this is an incorrect assumption in the case of high-bypass engines for civil aircraft. Typically, the bypass stream generated by the large fan, which can be over 3.5 metres in diameter, contributes up to 90% of the engine thrust. Hence, much of the power generated by the engine core is used to drive the fan. The engine is configured such that both the bypass stream and the hot-core jet issue at a relatively "modest" velocity not much higher than the maximum cruise speed of the aircraft (the Mach number

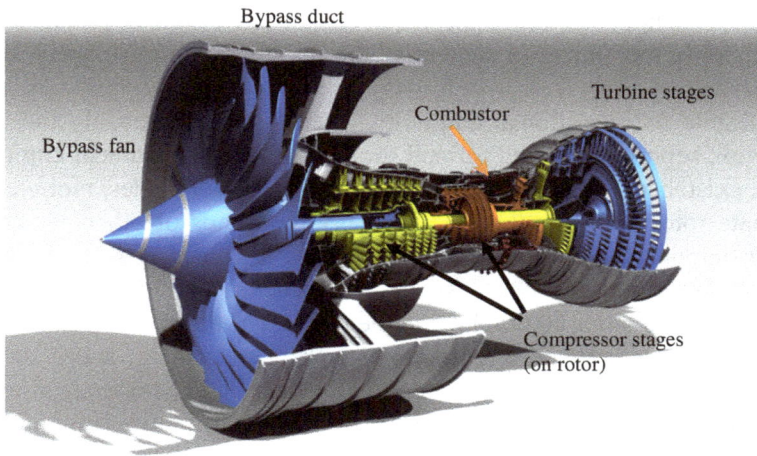

Figure 6.2. Rolls-Royce three-spool Trent 900 turbofan engine for large civil airliners. *Source*: C. Shakal, GrabCad, Stratasys Inc. (with permission).

being of order 0.85). This reflects the fact that the maximum propulsive efficiency is achieved by maximising the amount of air and gas discharged by the engine at a speed close to the speed of sound (Mach ~ 0.95).

The engine shown in Figure 6.1 contains three concentric spools, each connecting one or more turbine rotors to a set of compressor rotors (coloured yellow or red) or the fan (coloured blue). This allows different parts of the engine to operate at different rotational speeds, designed to optimise the engine's performance and maximise its efficiency. For example, the fan is designed to operate at a much lower speed than the high-pressure compressor/turbine assembly (coloured red).

Fighter-jet engines do not, as a rule, contain a bypass, and in this case, the thrust is entirely produced by the hot stream from the core. As this stream contains much of the energy released in the combustion chamber, the speed of the jet discharged from the nozzle is extremely high, as is desirable for a military jet. To increase the power and speed further, secondary combustion in an after-burner is used to maximise the thrust, especially during take-off. It is this secondary combustion behind the exit of the gas from the turbine that gives rise to the spectacular fiery tail behind a jet fighter during take-off in darkness. As an aside, it is interesting to remark here that the high speed of the hot-core jet was one reason why the jet engine, recognised as a means of aeroplane propulsion as early as 1920, was not used until the end of World War II. Early wooden aircraft were simply too flimsy to accommodate the power and gas speed of jet engines.

Despite its operational simplicity, a jet engine is probably the most complex, sophisticated and refined product of aero-mechanical engineering in existence. Large jet engines contain up to 5000 blades, mounted on alternate rotor and stator discs, and many thousands of other parts. The power output per volume is the largest of any engineering system, except for a rocket motor and possibly a nuclear reactor. The maximum engine power is of the order 60,000 hp and the maximum thrust is of the order of 25–35 tons. A single turbine blade can produce 250 hp, and a turbine wheel may contain more than a hundred blades. Engines operate at maximum temperatures of about 1800°K (higher in the combustor). Astonishingly, this value is well above the melting temperature of even the most exotic alloys and ceramic coatings used. Operation at this temperature is only possible because of spectacularly elaborate blade-internal and external cooling systems, as is shown in Figure 6.3. The exit jet normally issues noisily at close-to-sonic speed from the exhaust, while the jet-exit velocity of military jets, often accelerated by an afterburner downstream of the engine itself, can reach 3 times the speed of sound, discharging in an ear-splitting shock pattern, such as that shown in Figure 2.2. The gas stream within the engine is mostly close to sonic speed and is locally

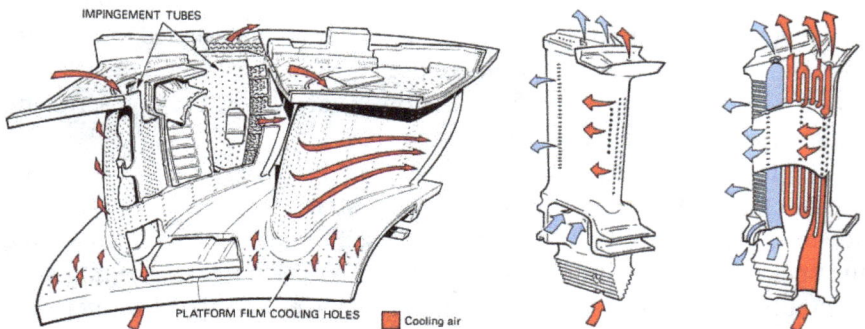

Figure 6.3.   External and internal cooling of turbine blades directly after the combustion chamber. Blades are hollow, rotate at high speed and carry cooling air in curved channels, while injection from within the blades through holes on the surface cools the outside of the blades.

Source: *The Jet Engine*, Rolls-Royce plc. (with permission).

supersonic. Rotors rotate at up to 15,000 rpm, and the blade-tip speed of rotation reaches supersonic values. Many components are subjected to enormous mechanical and thermal stresses, but they can operate safely for many hundreds of hours. This extreme complexity and the safety-critical aspects of large civil aircraft engines require technical expertise that only a few countries (USA, UK, Germany, France and Japan) have acquired over many decades of research and development. As noted already, by reference to Figure 6.2, modern high-bypass engines include several concentric spools (rotors), connecting different parts of the compressor to different parts of the turbine, each spool rotating at a speed different from that of other spools, so as to maximise operational performance and efficiency.

Although a blade may be viewed as a particular form of a wing, there are important differences between wings and turbomachine blades. The latter operate in a confined, rotating, hot environment, are twisted and bound closed flow passages along which the air or hot gas flows. The blades are mounted on the periphery of rotor discs, which are separated by stationary stator stages. The latter also contain blades, the task of which is to redirect the flow between sequential rotor discs. Yet, despite these major differences relative to wings, there are also some common aerodynamic characteristics. In particular, any blade contains a suction side and a pressure side, both curved in a manner similar to a wing profile. We shall look at the details of the flow around compressor and turbine blades shortly.

As seen in Figure 6.2, the compressor contains many stages, while the turbine only contains a few. The reason is that compressing air has to be done judiciously in modest increments because blades do not like operating in an increasing-pressure environment and may "stall" — a consequence of separation from the suction side, which has been described in detail in relation to wings subjected to an adverse pressure gradient at a large angle of attack (Section 4.5). Stall in the compressor passage is an ill-conditioned process that leads to a dangerous constriction in the effective flow area through the passage and thus a serious loss in compressor performance. Hence, any one compressor stage can only increase the pressure by a relatively small amount — at a pressure ratio of 1.2 or less. In contrast, blades easily tolerate operation in a declining-pressure environment, as is the case in the turbine within which the gas expands as the pressure falls, and this means that a large pressure drop is acceptable to any one turbine stage.

Figure 6.2 also shows that the flow passage contracts in the compressor parts of the engine core, up to the combustion chamber, while it expands in the turbine parts. As the mass flow through the core is constant (except for the minor addition of fuel in the combustion chamber), the flow area is indirectly proportional to the fluid density and the average flow speed. Conventionally, the average speed from one stage to the next is designed to be roughly constant. This must mean that the flow area must decrease with increasing density in the compression phase, while it must increase with decreasing density in the expansion phase within the turbine.

The manner in which the air flows through an "unrolled" compressor stage and the gas flows through a turbine stage is conveyed schematically by Figures 6.4 and 6.5, respectively. As can now be recognised, the blade sections are not dissimilar to the aerofoils discussed earlier, although the curvature of the turbine blades is higher. A steam turbine operates in a very similar fashion, although it is generally bigger and provides more

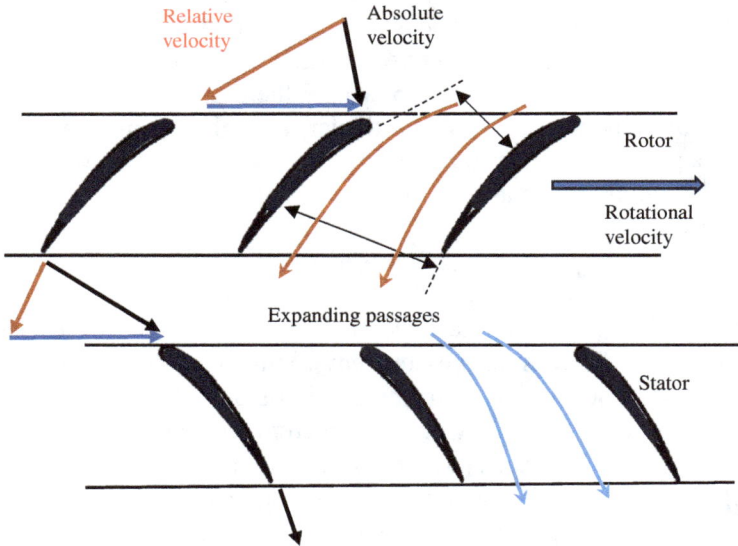

Figure 6.4. Schematic of the flow through an unwrapped circumferential surface cut across the blades of a compressor (rotor + stator) stage. The red arrows into and out of the rotor passages represent the directions of the flows relative to the passages. The actual ("absolute") flow velocity is a combination of the relative velocity and the rotational velocity, i.e., the black arrows. Note the expanding compressor passages, implying deceleration and compression.

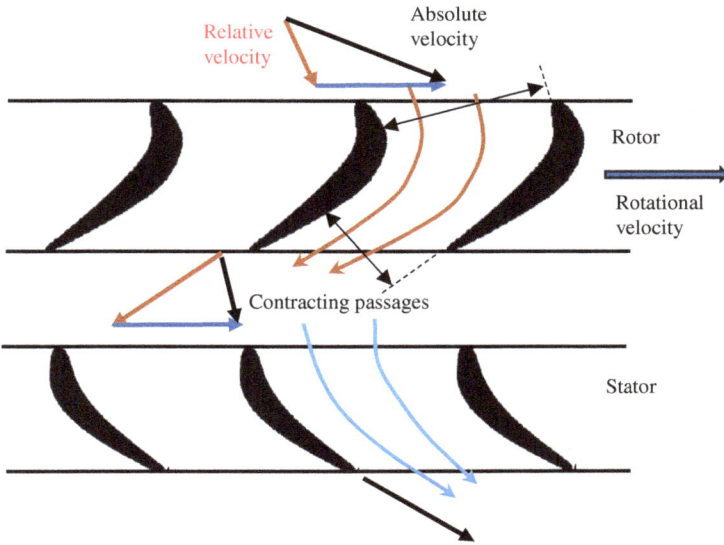

Figure 6.5.   Schematic of the flow through an unwrapped circumferential surface cut across the blades of a turbine (rotor + stator) stage. The red arrows into and out of the rotor passages represent the directions of the flows relative to the passages. The actual ("absolute") flow velocity is a combination of the relative velocity and the rotational velocity, i.e., the black arrows. Note the contracting turbine passages, implying acceleration and expansion.

power, and the geometry of its blades and passages is similar to those in a gas turbine.

The task of a compressor stage (Figure 6.4) is to drive the fluid into a region of higher pressure, thus increasing its energy, and to do so such that the absolute velocities at the inlet and outlet are broadly the same. The turbine (Figure 6.5) does the opposite: it allows the fluid to move into a region of lower pressure while keeping the absolute velocity across the stage broadly constant and reducing its energy by an increment that is transferred to the turbine rotor. In both stages, the task of the stators is to realign the flow from the exit of the rotor to the original inflow direction fit for the next rotor inlet.

Readers who do not wish to delve into the details of the flow path through the stages and the "velocity triangles" shown in Figures 6.4 and 6.5 might benefit from merely taking away the following relatively simple arguments.

The flow along the compressor-rotor passage, represented by the red arrows, is realigned towards the vertical (axial) direction and is decelerating (i.e., compressed), while the flow along the turbine passage is aligned away from the vertical to the horizontal direction, against the direction of rotation, and is accelerating (i.e., expanding). The realignment in the compressor case increases the flow momentum in the direction of rotation, and this is due to the compressor rotor pushing the fluid by its forward motion. Simultaneously, the pressure is rising because of the deceleration. Both increase the energy of the fluid, which is the objective of compression. In the case of the turbine rotor, the reorientation of the velocity (red arrows, again) and the increase in backward momentum causes the rotor to be driven forward. Simultaneously, the flow accelerates and expands, thus losing pressure and energy, the latter transmitted to the rotor.

The above explanation already covers part of what needs to be said about the "velocity triangles" in Figures 6.4 and 6.5. Again, the red arrows show the velocity relative to the blades. The actual — absolute — velocity of the fluid is represented by the black arrows and is derived by combining the relative velocity with the rotational velocity, represented by the blue arrows. So, the red arrows would be seen by an observer riding on the blades, while the absolute velocity would be seen by a stationary observer. The absolute velocity indicates the amount of the kinetic energy of the flow. This energy is falling in the turbine case, a decrease that indicates the energy transferred to the rotor. In contrast, the absolute velocity on the compressor case increases, indicating a transmission of energy from the rotor to the flow.

The combustion chamber has, usually, the form of a ring encompassing the circumference of the engine, but there are also some engines in which combustion occurs in an array of can (tubular) combustors distributed around the circumference of the engine. Air enters the chamber from the compressor, the fuel is injected through a circumferential array of injectors, and the mixture is ignited, reaching temperatures of the order 1600°C. Due to this very high temperature, the combustor walls have to be cooled. This is done by injecting some of the air delivered by the compressor from outside through holes in the combustor walls into the outer part of the reaction zone. As explained earlier by reference to Figure 6.3, the hot gas enters the turbine at a temperature that is higher than the melting point of the turbine-blade metal, and this requires an elaborate blade-cooling system, as shown in the figure. After the expansion in all turbine stages, the gas exits the core at a temperature of around 600°C.

In engines that drive a propeller, a helicopter rotor or a generator, the last few turbine stages — being the largest and extracting most of the energy from the hot gas stream — are connected to the driven device, instead of the bypass fan. This absorbs the difference between the total power produced by the engine and the power needed to drive the compressor. In these applications, the objective is to minimise the gas-exit energy, because this is not used for propulsion, and to extract the maximum amount of energy from the hot gas within the turbine to drive the propeller or rotor.

6.2 Power to My Blades: Wind Turbines

The history of windmills goes back to Babylonian times, in the second millennium BC, and these were used extensively for milling until the 19th century. Of interest here, however, are modern wind turbines that started gaining popularity for power generation in the early parts of the 20th century. There are now hundreds of thousands — maybe half a million — wind turbines used worldwide, many for local low-power applications in rural and densely built-up urban settings, but increasingly for feeding national grids with electricity to replace fossil-fuel-based generation.

There are several types of wind turbines, some basic forms shown in Figure 6.6. Vertical turbines are relatively small, and their use is mostly limited to local low-power generation. The horizontal 3-bladed type is by far the most important one, and it strongly dominates the wind-energy sector. Typically, such turbines are erected in wind farms containing dozens of turbines, as shown in Figure 6.7.

The quest for ever-greater energy yield over the past few years has driven wind turbines to astonishing dimensions, the largest configurations being installed in off-shore farms. The largest single blade is 130 metres long — see Figure 6.8 — and the tallest turbines are around 300 metres high. In favourable wind conditions, the largest turbine can generate around 10 MW. This output suffices to supply a town of 100'000 households, but it must be born in mind that this output is far from continuous and will be much lower in calm atmospheric conditions. In fact, the power derived can be shown to be of the order

$$power = 0.5 \times density \times area\ swept\ by\ blades \times (wind\ speed)^3$$

So, the output is very sensitive to the wind speed, due to the cubic dependence. This is one reason why it is desirable to mount the rotor as high up

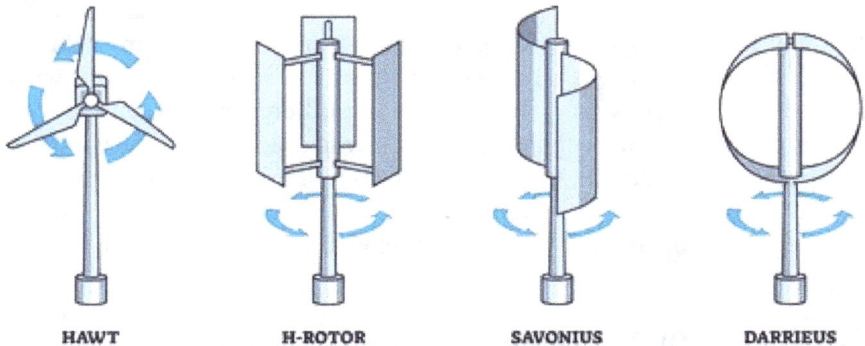

HAWT   **H-ROTOR**   **SAVONIUS**   **DARRIEUS**

Figure 6.6.   Different types of wind turbines.
Source: iStock (with permission).

Figure 6.7.   A wind farm of horizontal-axis wind turbines.
Source: iStock (with permission).

from the ground as possible where the wind is less slowed down by ground friction. Even at optimal conditions, the turbine output is modest when compared to an output of around 1 GW derived from a nuclear power plant — i.e., one hundred times higher than the turbine output — while a typical gas-powered station generates around 200 MW. Hence,

Figure 6.8.   Images conveying the huge size of modern horizontal-axis wind turbines: (a) mould of a 115m long blade having a 3m leading-edge-to-trailing-edge span; (b) a 90m long blade being road-transported.

Source: Siemens-Gamesa (with permission); Mammoet Transport (with permission).

even if we ignore the intermittent nature of the yield from a wind turbine — the average output being around 30% of the maximum — we need around 50–100 wind turbines to replace a large conventional power plant, and this is the reason for the installation of large off-shore wind farms in which the turbines are larger and where the wind is stronger and more sustained.

Some readers will be curious about why horizontal turbines only contain a very small number of thin blades, while old windmills have many-bladed rotors with large blades which cover a large proportion of the area within the diameter of the rotor. You might also wonder why it is that a wind turbine looks so different from a gas-engine turbine, which might contain 70+ blades The easiest way to answer this question is to say that modern turbine configurations are the result of many years of small-step optimisations that targeted an optimum compromise between torque, drag, rotational speed, power output, standard wind conditions and structural constraints — weight, material properties, strength, flexibility and acoustic properties (noise). A gas turbine is confined by an outer casing, its blades forming expanding flow channels, operates in a hot-gas environment that is compressible, rotates at 10–15000 rpm and provides a huge amount of power in a small volume. A wind turbine, in contrast, operates in gentle wind, rotates slowly, is not confined and relies on its size to provide a meaningful power output. These are very different operational scenarios for which different types of turbines are optimised. The same applies to compressors, fans, propellers and helicopter rotors, which have very different geometries yet are essentially doing the same thing — imparting

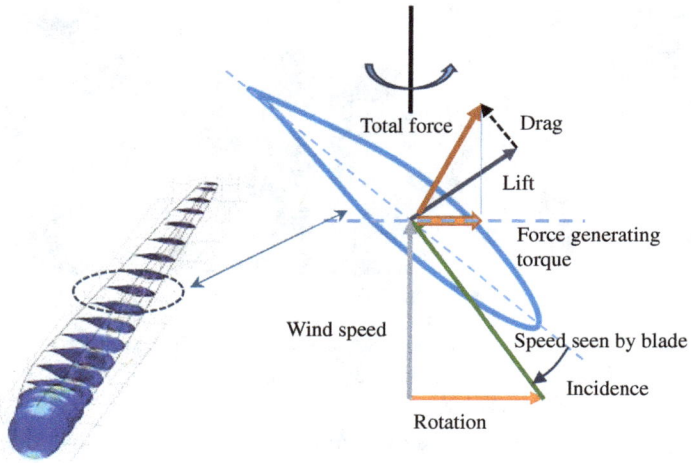

Figure 6.9.   Typical variation in the blade profile along the blade and the aerodynamics by which oncoming wind generates torque on the blade.

energy to a fluid and pushing it along. We shall deal with propellers in the following section.

However, it is also possible to give a more explicit answer to the question on the blade configuration of horizontal turbines. We shall do so in the following description, but we look first, in Figure 6.9, at the basic aerodynamics at play — specifically, how a blade responds to the wind. The interaction might look complicated, but a little perseverance, supported by the explanation below, will readily make it clear.

Consider a cut through a blade pointing out of the page from the axis of the turbine. The blade is moving to the right, rotating in the clockwise direction on a vertical shaft that lies somewhere below the page surface. The wind is indicated by the blue wind-speed arrow. This speed is the absolute velocity, seen by a stationary observer, while the speed seen by the blade, the green arrow, is a relative velocity (the difference between the two was also explained by reference to Figure 6.5 for the gas-engine turbine). The latter has an angle of incidence relative to the blade, and the result is a lift force, the black arrow, which is normal to this direction, while the drag force is parallel to this direction, the dashed black arrow. The total force is the sum of the two, the long red arrow. Finally, the force driving the blade in the direction of rotation is the thick red arrow, which is the projection of this total force onto the horizontal, the direction of rotation. This is how the wind is able to drive the rotor in motion.

It is obvious that we want to maximise the lift and minimise the drag, for this will increase the torque-producing force. Also, we do not want our rotor to move too fast relative to the wind speed, otherwise the angle of incidence in Figure 6.9 will decrease and may even reverse. A typical turbine does not rotate at a high speed, and this is also conducive to the noise level. As the rotational speed increases linearly with the radius, the speed seen by the blade (the green arrow) will tend to twist towards the horizontal direction, if the wind speed is constant across the rotor. Hence, to maintain the angle of incidence and the force configuration shown in Figure 6.9, we need to twist the blade as we move from the hub to the blade tip, and this twist is seen in any horizontal wind turbine, as it is, indeed, seen in almost all rotating-blade configurations, including propellers, jet-engine blades and helicopter rotors. Typically, the rotational speed is chosen such that the blade tips are moving at no more than 7–10 times the wind speed.

We return to the question on the number of blades and their slenderness. The answer is a combination of several factors, some fairly obvious and others reflecting experience gained over many years. First, we wish to reduce the number of blades to a minimum on structural and cost grounds. Large two-blade rotors tend to wobble and are structurally unsafe. Three-bladed rotors are more stable and dynamically balanced. A high number of blades makers the rotor heavy and expensive to manufacture and maintain. Increasing the number of blades may result in a larger torque, allowing higher rotational speed and higher power yield, but an increase in speed increases noise, erosion and mechanical stresses, reduces efficiency and requires higher twist to maintain an acceptable angle of attack along the blade (see Figure 6.9). The blades have to be thin in order to keep their drag and weight low; both rise rapidly with the span of the blade profile (the leading-edge-to-trailing-edge dimension). Thin and radially long blades reduce spanwise variations in the flow around the profile and enhance two-dimensional flow conditions along the blades. The area covered by the blades, within the swept area, should be kept low, so as to reduce blade-to-blade interference. Increased blockage causes radial deflection of the oncoming wind, thus reducing the flow through the swept rotor-disc area and distorting the flow around the blades by increasing the radial-flow component along the blades. Blockage also lowers the velocity in the wake behind the turbine and, thus adversely affecting the output of downstream turbines within a wind farm. A very large radius is desirable, as this exploits the wind over the largest possible swept area. This series of statements should suffice to make it clear that the shape and configurations of a wind turbine are the outcome of a multi-parameter optimisation, involving aerodynamics, structural mechanics and operational factors.

Finally, because the wind conditions generally vary greatly in time and space, with the power output depending cubically (to the power of 3), a wind turbine needs to be configured so as to function acceptably at low as well as high wind speeds. This can be done by active control of the blade orientation relative to the wind. For example, in stormy conditions, the turbine blades have to be protected, both aerodynamically and structurally, and this can be done by pitching the blades away from the wind, so as to cause the blades to stall, thus reducing the lift and the torque on the rotor.

6.3 Push and Pull Me: Propellers

A propeller does the opposite of what a turbine does. A turbine reduces the momentum and kinetic energy of an oncoming fluid stream and converts this energy into useful rotational work. A propeller is driven by mechanical work and increases the momentum and energy of the fluid passing through the circular area covered by the propeller blades. The consequence of the increase in fluid momentum is a force on the propeller, expressed by a relationship stated earlier (Section 4.2, Figure 4.7) in the context of the thrust of a jet engine. It is repeated below:

$$thrust = m_t \times (V_j - V)$$

where V is the speed of the air approaching the propeller and V_j, previously the speed behind the jet engine, is here the higher speed behind the propeller. To a first approximation, the mass-flow rate, m_t, is

$$m_t = density \times (area\ swept\ by\ propeller) \times V_j$$

So, we see that the thrust depends quadratically on the speed. Clearly, the primary operational output of a propeller is not the increase in flow momentum but the generation of thrust that is transmitted to the body on which the propeller is mounted. This is in contrast to the objective of a cooling fan — e.g., a room ventilator — for which the desired quantity is the speed behind the fan, V_j.

Although we are not dealing here with pumps, it is instructive to point out that a rotating pump impeller, such as the compressor in the jet engine described in the previous section, is closely related to a propeller. The main difference is that a pump transfers the power input mainly into pressure energy rather than kinetic energy. In many applications, pumps have a single task: to increase the pressure in a flow at a near-constant speed. This is so in a central-heating pump, for example, which transports warm water through a pipe of constant cross-sectional area. It has to increase the pressure of the water to overcome the frictional resistance of the long pipes and radiators in the central-heating system. It is recalled that Section 4.2 contains an explanation as to why pressure is a form of energy. This is most obvious when a pump is used to transport water from a lower-level reservoir to a high-level one.

As explained in Section 6.1, by reference to Figure 6.4, the relative-to-passage velocity in the rotating compressor (the red arrows) is decreasing, suggesting that the pressure is increasing. At the same time, the absolute velocity (black arrows) is increasing, signifying that the kinetic energy is increasing. Hence, our compressor wheel is a kind of hybrid between a pure pump and a pure propeller, in so far it increases the pressure and the kinetic energy simultaneously. Another important difference between our compressor and a propeller is that the former is ducted and contains many blades, typically 20. An open propeller, on the other hand — be it an aircraft propeller, a ship propeller or a helicopter rotor — contains a much smaller number of blades, typically 3–5.

While the basic operational purpose of any propeller is common to all propeller forms, they differ greatly in terms of their detailed design, especially the shape of the blades, as is illustrated by a few examples included in Figure 6.10. In fact, there are substantial geometric differences and variations within any one of the groups included in the figure, as is exemplified by two out of many aircraft- and ship-propeller geometries used in practice, and much the same can be said about the other groups.

Whatever its precise shape, a propeller blade has much in common with a wing, in principle. Except for very basic low-cost/low-performance applications, the blade has a cross-section similar to that of a wing, having a suction and a pressure side, and it is moving relative to ambient fluid

Figure 6.10.   Examples illustrating the wide variations in open (unconfined) propeller geometries. Upper row: aircraft propellers and helicopter rotor; lower row: ship propellers and cooling-fan rotor.

Source: Wikimedia Commons CC BY-SA; iStock, upper left and upper right images (with permission).

body in a manner similar to that of a wing on an aircraft in flight. The basic difference is that the wing-equivalent flow of air relative to the blade is created by the rotation of the blade.

The manner in which a propeller blade generates thrust is explained in Figure 6.11. In common with the wind-turbine configuration in Figure 6.9, the blade is assumed to be mounted on a vertical shaft somewhere below the page surface. Here, however, it rotates in the anti-clockwise direction, moving to the left. On the left of Figure 6.11 are shown the components of motion of the blade due to rotation and forward motion, the green arrow being the resultant speed. When this arrow is turned by 180°, it represents the speed seen by an observer sitting on the blade. Due to the angle of incidence, the blade generates lift normal to its chord, and there is a drag force parallel to the chord, giving rise to the total force on the blade, represented by the thick orange arrow. This can finally be decomposed into a component in the direction of rotation and normal to it. The latter, represented by the thick green arrow, is the thrust that pushes the blade and aircraft forward.

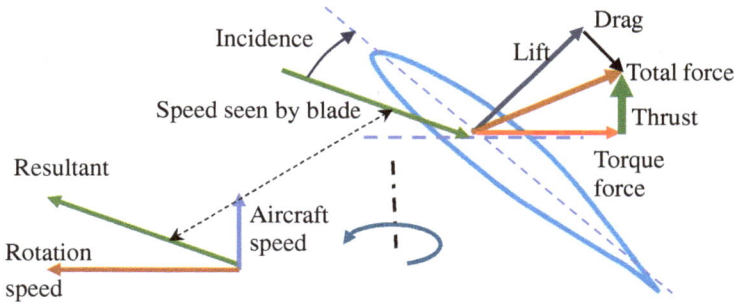

Figure 6.11. The aerodynamics of a propeller profile in the presence of forward motion. The profile is a cut through a blade rotating in anti-clockwise direction about a vertical axis assumed to be located somewhere below the profile (into the page).

As in the case of the wind turbine, the speed of rotation varies linearly with the radius along the blade, being a minimum at the root of the maximum and maximum at the tip. Hence, as before, to maintain a constant angle of attack, the blade has to twist. This is a feature common to all radial blades mounted on propellors. In addition to the twist, aircraft propellers often include gear mechanisms in the hub that allow the blades to be twisted, so as to control the blade's angle of attack. In fact, the thrust on short-haul passenger aircraft can be reversed by twisting the blades in such a way that they either stall or even generate a force opposing the forward motion of the aircraft, so as to assist its deceleration.

There are many reasons for the differences in shape shown in Figure 6.10. Most importantly, the design of the blades greatly depends on the specific operational objectives, the fluid properties (especially the density), the mass flow rate of fluid to be propelled to achieve the requisite thrust, geometric constraints imposed by the vehicle to be pushed or pulled or the device accommodating the propeller, the speed of rotation, weight and structural-strength limitations, acoustic characteristics (e.g., in submarine applications and domestic fans) and the cost of producing the propeller. Thus, like most other rotating machinery, the design of propellers is subject to a complex multi-parameter optimisation that aims to reconcile many constraints with the highest possible efficiency and propeller performance.

Aircraft propellers and helicopter rotors tend to have slim and long blades. They have to rotate at high speed to propel low-density air at a high mass flow rate. They need to be light and withstand extremely high radial forces and aerodynamic and structural distortions, arising from rapid changes in the direction of flight. This favours a low number of slim blades along which pressure variations and thus propulsive performance remain high over most of the blade length. A low number of blades is also conducive to efficiency, as it reduces blade-to-blade interference by the wake of any one blade engulfing the blade trailing it. On the other hand, a larger number of blades allows higher thrust to be achieved at a lower propeller or rotor diameter for a given rotational speed, and it is now common to see rotors with 6 and even 8 blades on large military transport planes (e.g., the Airbus A400M aircraft), thanks to recent developments in material science that allow light, high-strength composite blades, combining metals, carbon fibres and plastics, to be manufactured.

A helicopter rotor is an extreme form of a propeller, in terms of its geometry, its elaborate mounting on the rotor axis and the complexity of the aerodynamics within its plane of rotation. The rotor blades need to be especially long to create a large-diameter downwash, required to lift the helicopter and its heavy load. As with any propeller, the thrust of a rotor is proportional to the square of the tip speed. A high speed is thus conducive to the thrust, and the blade-tip Mach number is designed to reach a value of 0.7–0.8. At this value, the flow over the rotor suction side can be transonic, i.e., the Mach number exceeds 1 over the front of the rotor blade and reverts to subsonic speed towards its rear through a weak shock. This is the reason why we occasionally hear a rapid staccato of popping sounds when a large helicopter flies overhead.

A low rotor blade area is desirable to achieve low drag. Fewer blades benefit low weight and reduced inter-blade wake and blade-tip-vortex interference within the plane of rotation, but this choice needs to be compensated by a high blade length so as to achieve the required thrust. The blades need to be very light and flexible and need to be controllable in terms of their pitch, so as to adapt their aerodynamic characteristics to highly variable flight conditions — e.g., hover, fast forward flight and rapid change in the orientation of the axis of rotation. Finally, careful attention has to be paid to the load distribution over the length of the blade, so as to avoid a high load towards the tip of the blades and thus an excessive bending moment at the root of the blade. This constraint is met by an appropriate choice of the blade profile, the angle of attack, the twist and the taper towards the blade tip.

A peculiar operational feature of a helicopter rotor is that it has to provide sufficient thrust within a broad range of forward-flight speed in the direction almost normal to the rotor's axis of rotation. The addition of the adverb "almost" signifies the fact the rotor has to be tilted relative to the flight direction, so as to create a forward-trust component, which propels the helicopter horizontally. A consequence of forward flight is that the blades moving in the flight direction are exposed to an air velocity that is the sum of the rotational component and the flight speed, thus producing extra lift, while the reverse applies to the receding blade. Apart from rendering the aerodynamic performance of any one blade highly time-dependent, the thrust produced by the rotor acts on a line that is displaced relative to the rotor axis. This asymmetry results in a torque on the helicopter that needs to be compensated by a slight transverse tilting of the rotor or/and adjustments to the angle of attack by dynamically twisting the blades.

As seen in Figure 6.10, the shape of ship propellers differs dramatically from that of aircraft propellers. In contrast to the latter, the former rotate slowly, need to be radially compact (close to the hull) and are designed to propel a very large mass to produce the high thrust required to drive ships that are subjected to very high drag even at moderate speed. The design of ship propellers is, again, a complex optimisation exercise, involving a reconciliation of a number of contradictory constraints, as is exemplified by the following juxtapositions.

A low propellor speed reduces drag and is conducive to efficiency. However, this needs to be compensated by a large propeller diameter, which is constrained by the hull as well as blade-strength limitations. Also, the propeller rotation frequency must be different from the resonant frequencies of the shaft, hull and engine mountings. Thin blades with low blade area tend to benefit efficiency and the uninhibited flow of water between the blades, but this can compromise the structural integrity of the blades. The number of blades can be increased to reduce the load on the individual blades, but this requires care to avoid vibrational coupling among the blades. A high angle of attack and blade curvature (the "camber") favours high thrust, but this can easily cause a phenomenon called "cavitation" on the low-pressure suction side of ship-propeller blades.

Cavitation is an especially important factor in the mix of considerations affecting design choices for ship propellers. This phenomenon is, essentially, a localised boiling of water at low temperatures due to the

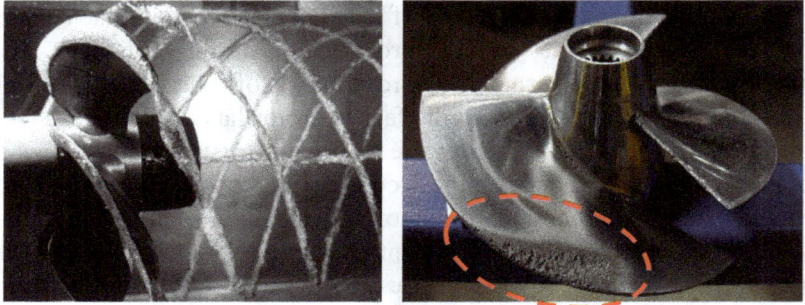

Figure 6.12. Cavitation bubbles produced by a rapidly rotating ship propeller and the material damage that is caused by imploding vapour bubbles.

Source: Wikimedia Commons CC-BY.

pressure falling below the vapour-pressure limit. For example, at a temperature of 20°C, water boils at a pressure of approximately 2–3% of the sea-level atmospheric pressure. Thus, if the pressure on the suction side is lower than this limit, the water next to the blade evaporates in the form of bubbles. When these bubbles move into regions of higher pressure across the blade, they collapse (implode), and this collapse generates explosive pressure waves that tend to cause the type of severe damage to the material of the blade shown in Figure 6.12. This process is also extremely noisy, as is occasionally perceived in heated pipe flow when this is highly constricted by a nearly closed valve acting against a high pressure. Cavitation is, therefore, a critical issue in submarine propellers.

6.4 A Noisy Beast

Noise has been mentioned repeatedly, albeit in passing, in the context of jet engines, wind turbines, ship propellers and the buzzing or humming that arises due to wind interacting with wires and buildings (see Figures 4.30, 4.31 and 4.33). Here, we expand on the subject to make the point that almost all flows, especially turbulent ones, cause noise, or at least acoustic disturbances. We might not like it, but we have to live with it.

We perceive noise (and any other sound) as an interpretation of pressure fluctuations on our eardrums. So, anything that causes pressure fluctuations also causes sound and noise — as, for example, the puffs discharged from motor-vehicle exhaust pipes. Noise also occurs in solid, of course, when two solid bodies impact on, or rub against, each other. This generates vibrations in the solid, which are then transmitted as sound waves through the air. Explosions aside, some fairly ordinary fluid flows are capable of generating extreme noise levels that can easily cause damage to our ears. There are three main sources: vortices, turbulence and shock waves at supersonic speeds. The first two may be viewed together, as turbulence is essentially a collection of vortices.

Any turbulent flow generates noise, not all perceived as unpleasant. For example, most of us will regard the sound emanating from a shallow stream or river water flowing over a partially exposed stony bed, or from gently breaking waves on a seashore, as soothing. As explained in Chapter 5, turbulence is a collection of eddies. These move, change in size, collide, break down, coalesce and generally deform rapidly in space and time. The collision of fluid packets within the eddies, in particular, causes pressure peaks. We also know that vortices feature pressure variations within them, with minima at their centres. Therefore, the unsteady evolution of the turbulent stream is the cause of pressure fluctuations and hence noise. These phenomena are common to all turbulent flows that are subjected to shear, be they free jets, or near-wall boundary layers, in which case noise is also generated by eddies shearing against, and impinging on, the solid walls.

We are familiar with one particular source of intense noise — the jet engines on a civilian plane, shown in Figure 6.13 — a visualisation of a computational simulation. The flow approaching the exit nozzle is already highly turbulent because it is being churned around by the turbine stages of the engine. Turbulence is substantially increased by the jet exiting the nozzle at high speed, shearing against the nozzle walls and the much slower ambient air. Similarly, the discerning (or anxious) flier will be aware that the extension of the landing gear and the deployment of slats and flaps, the latter to increase lift during landing at low speed, increase the level of noise and vibrations in the aircraft, as turbulence is generated by the action of shear layers formed on the various devices and the links and wheels within the landing gear. This is exemplified by the results of computations simulations in Figure 6.14.

Figures 6.13 and 6.14 not only show the turbulence fields but also give visual information on how the acoustic waves propagate away from the turbulent shear layer. In the case of the landing gear, we also see intense waves upstream of the landing gear, which are created by the pressure rise and fluctuations in the forward-facing impingement area around the wheels.

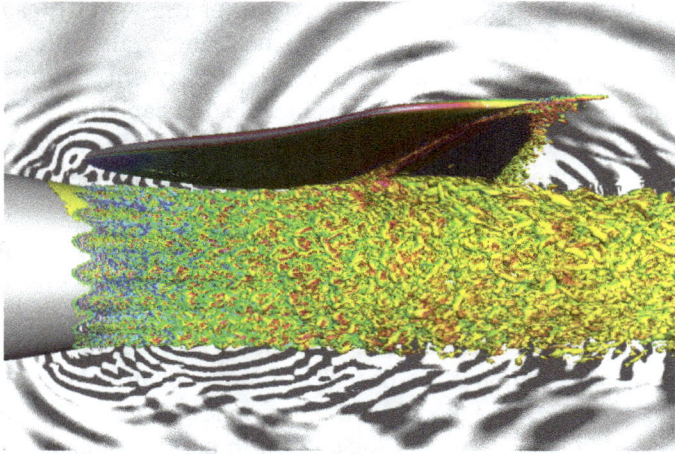

Figure 6.13. Turbulence in, and noise emanating from, a jet discharged from a serrated jet-engine nozzle below a wing.

Source: Wang *et al.* [29] (with permission).

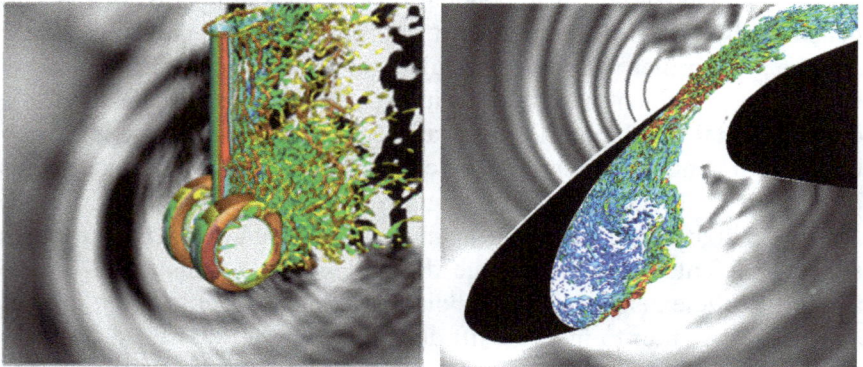

Figure 6.14. Turbulence behind, and noise emanating from, a landing gear and a slat extended from the main element of an aircraft wing.

Source: Redonnet *et al.* [30] (with permission); T. Knacke and C. Mockett, Upstream CFD (with permission).

An intriguing feature of recent civil aircraft jet-engine-exit nozzles is the serrated geometry of the exit lip, shown in Figure 6.13. This design arises from the observation that the acoustic signature of a suddenly imposed intense shear at an edge is reduced if the edge has an irregular geometry. In essence, the irregularity prevents the formation of long coherent vortices along the separation line, which reduces the intensity of the sound emitted by the turbulent layer downstream of the edge. Serrations of the type shown in Figure 6.13 are also used, in the form of serrated extensions, along the trailing edges of some large wind-turbine blades, especially on turbines located close to populated areas, in an effort to reduce the noise emanating from the blades. While the reduction in noise is fairly modest — of the order of 3 dB — this level is deemed to be of sufficient environmental impact to justify the additional construction cost. An incidental advantage, reported in experimental studies of realistic blade configurations, is a modest increase in output power (around 1%) due to the increased lift on the blade arising from the added surface area of the serrated strip.

Shock waves that emanate from supersonic aircraft, rockets and jets are, essentially, structures across which intense pressure changes occur in a step-wise manner, i.e., they are discontinuities. We have already encountered two relevant cases much earlier in the text, namely, the rocket plume in Figure 2.1 and the supersonic jet in Figure 2.2 of Chapter 2. We perceive the pressure rise or drop across a shock wave as a sharp crack or a "boom". Supersonic jets issued by fighter jets and rocket engines are extremely noisy. Some more senior readers might recall the noise generated by the Concorde supersonic airliner taking off with its military-grade engines on full power. The turbulence within supersonic jets is a complex collection of subsonic and supersonic eddies, from which a huge number of shocklets emanate, often in the form of a painful screeching noise, as illustrated by the computational simulation in Figure 6.15. The left-hand image, in particular, shows that a wavy, oscillatory, behaviour of the shock

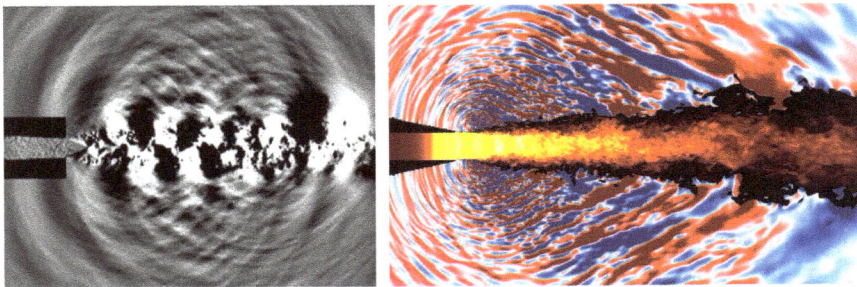

Figure 6.15. Turbulence in, and noise emanating from, supersonic "screeching" jets.
Source: J. Sesterhenn and J. Schulze (private communication); Chen *et al.* [31] (with permission).

structure plays an important role in the screeching noise, manifested by the emission of powerful periodic sound waves from the supersonic jet.

Hopefully, the short description given in this section will encourage you, the reader, to listen to the many types of noise around you with a greater degree of awareness, to try to figure out whether it emanates from a fluid flow and possibly identify what that flow is.

6.5 Swirl a Vortex

We have encountered vortices and eddies numerous times in the preceding chapters: first, in the context of weather systems, then when explaining the formation and properties of vortices and discussing vortex shedding and recirculation bubbles, and finally in relation to turbulence and eddies within a number of turbulent flows. In virtually all these flows, vortices formed as a consequence of geometric constraints or inevitable flow mechanisms that are not within our control. There are many applications in engineering, however, in which vortices are generated quite deliberately to achieve particular operational objectives. An indication of one technique was already given in Section 4.8: the tangential injection of fluid into a circular chamber, but no reasons were given as to why such action is useful. We consider a few related examples below.

Many readers will be familiar with the Dyson family of vacuum cleaners. Some may even be aware of the multi-tube "cyclone" that characterises these devices (and other vacuum cleaner brands as well). How do these devices work? To understand this, we focus first on a much larger device — the industrial cyclone separator shown in Figure 6.16. Variants of this separator can be up to 10 metres high and are used in the process, petrochemical, construction and food-processing industries to separate solid particles from gas–solid or liquid–solid mixtures, thus resulting in a clear or dilute gas or liquid stream and a dense particle stream. An exotic application is the separation of insects from air intended for ventilation or clean laboratory environments.

In Figure 6.16(a), a dusty stream enters the cyclone separator through one tangential port or several of them distributed circumferentially at the top of the separator. This tangential injection creates a strong vortex inside the cylindrical and conical sections, within which the pressure at the centre is low relative to the outside. By the action of centripetal acceleration

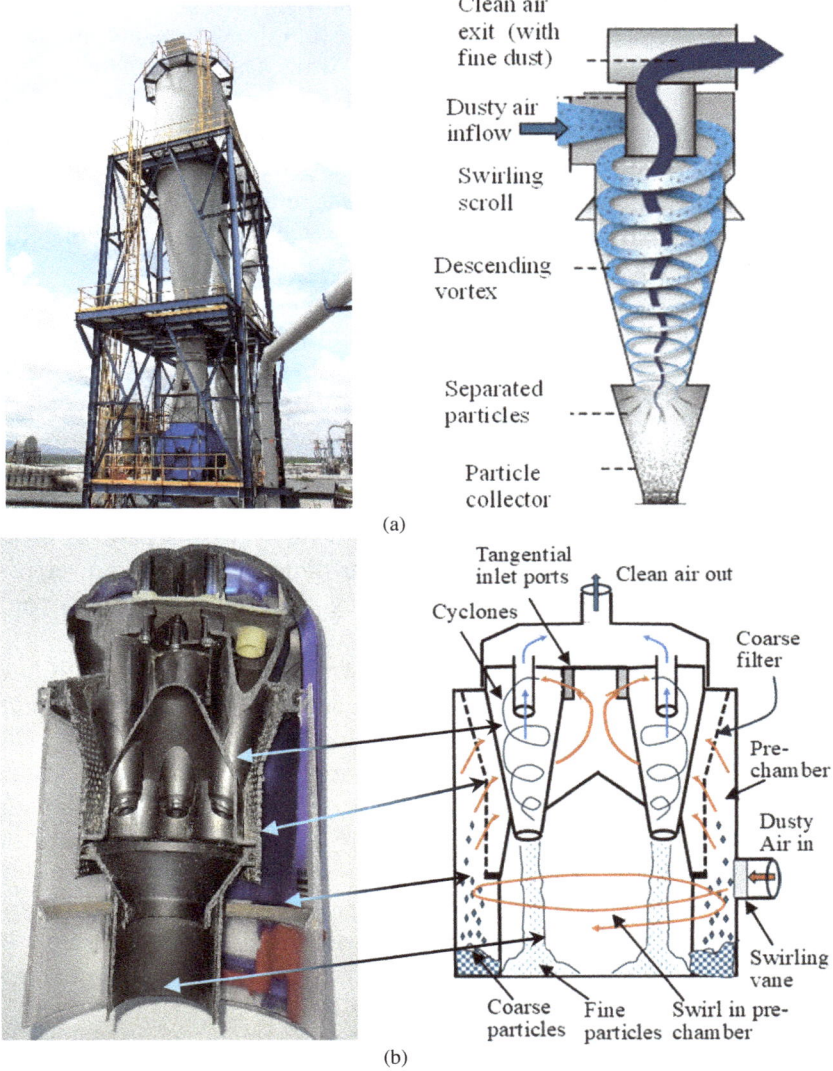

Figure 6.16. Cyclone separators in (a) an industrial facility and its operating principles and (b) the Dyson vacuum cleaner, with a simplified schematic indicating its operational principles.

Source: (a) Pharmacy Images, https://pharmacyimages.blogspot.com/.

and related centrifugal force, the heavier particles, which are not in equilibrium with the radial pressure variation within the vortex, are migrating outwards and move downward along the wall. The dilute carrier fluid, from which most particles have (ideally) been removed, is being sucked out from the centre of the vortex, and the mass extracted is being replenished by the downward-moving dusty stream. Towards the bottom, there is a diminishing amount of air left, and the tapered construction serves to amplify the remaining part of the vortex, as well as aiding the collection of the particle phase at the bottom.

As shown in Figure 6.16(b), the principle of the Dyson vacuum cleaner is, essentially, the same — although the construction details are more elaborate. Also, the cleaner drum has several of these separators to reconcile the small scale of any one separator drum with the large volume of air sucked by the cleaner through the system. The separator is only able to separate the large-to-medium-size dust particles from the dusty air, and the cleaner relies on fabric filters to remove the smallest particles. These particles are always the hardest to remove in a cyclone separator, as the centrifugal force declines with their linear dimension to the power of 3 (the mass), while the drag force resisting their outward motion only declines in proportion to the power of 2.

Uranium centrifuges of the type shown in Figure 6.17, which have featured repeatedly in the media in the context of the Iranian uranium enrichment programme, rest on the same operational principles as other rotating-flow separators. Although different in construction from the cyclone separators shown in Figure 6.16, they also exploit differential centrifugal forces to separate U235 from U238, the former being the desirable fissionable product. As shown in Figure 6.17, this is done in large cyclones with the outer drum rotating at up to 50000 rpm. As the two isotopes, processed in gaseous uranium hexafluoride form, differ only by about 1% in weight, this is a very slow process, in which the uranium gas is channelled through hundreds or even thousands of centrifuges in succession.

Another important application of intentionally generated vortices aims at securing stable and efficient combustion in many combustion chambers used in industry and aviation. The principle is explained with the help of Figure 6.18(a), while the realisation of the principle is illustrated in Figure 6.18(b), which shows a jet engine can combustor. Two streams enter the combustor: an outer air stream that is swirled in an annulus filled with curved stationary vanes, similar to blades, and an

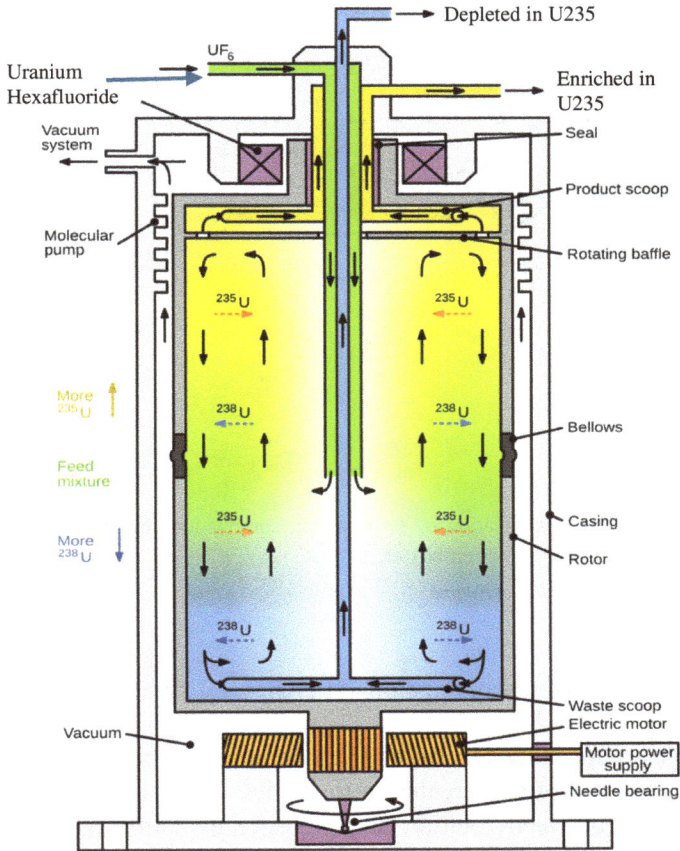

Figure 6.17. Centrifuge used to separate the Uranium isotope U235 from U238.
Source: Wikimedia Commons, CC0.

inner non-swirling fuel jet that is normally injected as a spray. To achieve combustion stability — i.e., to ensure that the flame is anchored and sustained and not simply blown off by the fast stream — hot products need to be transported backwards from the hot downstream region to the upstream region where the spray is injected. Flame stabilisation can be achieved, in principle, by placing a solid baffle within the stream, so as to provoke a recirculation bubble behind it, but this method is neither efficient nor operationally safe. A better technique is to exploit the relationship between swirl and pressure to achieve stable combustion, in

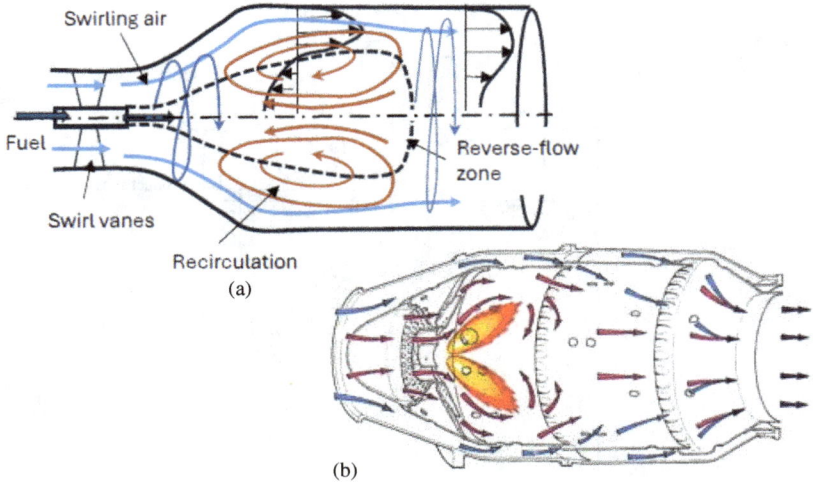

Figure 6.18.   Flame stabilisation by swirl within combustion chambers: (a) operational principle; (b) a jet-engine can combustor.

Source: (b) Rolls-Royce, *The Jet Engine*.

addition to increasing combustion efficiency by swirl-promoting turbulent mixing.

When the swirling air enters the combustion chamber, a region of intensely low pressure is created in the centre of the chamber, consistent with the link between rotation and pressure explained in Section 4.3. As the flow progresses along the combustor, it slows down within the divergent combustor part. At the same time, the swirl intensity declines, and the swirl momentum is redistributed due to the combined action of wall friction and turbulence-induced mixing. The combination of deceleration and decline in swirl result in a strong rise in pressure — or rather an axial recovery from the very low pressure — in the central region of the swirling stream. This pressure rise provokes a rapid slow down of the fluid in the central part of the flow, around the centreline, until fluid begins to flow backwards, thus transporting hot reactants upstream and igniting the fuel-spray/air mixture.

The final example in this section is the flow in the combustion chamber of an internal combustion engine, as shown in Figure 6.19. Here, the swirling motion is generated in a curved passage within the intake pipe. The objective of the swirl is here to promote mixing and efficient

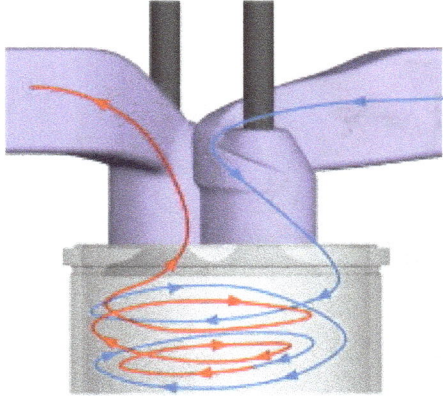

Figure 6.19.   Swirling flow in the cylinder of an internal combustion engine.

Source: Zheng Xu (private communication).

combustion once the air/fuel mixture is compressed by the piston, and combustion is initiated by the spark plug.

6.6 Blowing Hot and Cold

The ability of fluids to absorb, reject and transport heat over long distances is extremely important in many fluid-flow applications. We are familiar with central-heating boilers, radiators, heat pumps and air conditioning units increasing or decreasing the temperature in the built environment, and with car radiators designed to cool the engine's cooling water, but we are perhaps less aware of the relevance of heat transfer in heat exchanges, nuclear reactor channels and power-generation applications. Massive heat exchangers, like the one shown in Figure 6.20, are used, for example, in heating and cooling steam in steam-turbine cycles of power stations. One exotic heat-transfer application we have encountered already is the internal cooling of jet-engine turbine blades, as shown in Figure 6.3.

Heat transfer is driven by two main ingredients: a temperature difference — e.g., between two fluids separated by a solid wall — and the resulting flow of heat into or out of the two fluids by a combination of conduction, turbulent mixing and convection. Conductivity is, fundamentally, very similar to viscosity: the later mixes momentum by Brownian motion acting on a velocity difference across a shear region, while the

Figure 6.20.   A heat exchanger used in a steam-turbine power station.
Source: Titan Metal Fabricators (with permission).

former mixes heat, also by Brownian motion, across a variable-temperature layer. However, the conductivity and viscosity are not the same, not even in terms of their dimensional units, and there are significant differences between temperature conduction and viscous momentum mixing in different liquids. For example, liquid metals conduct heat far better than does water, although their viscosity values are fairly close.

It has been argued in Chapter 5 — e.g., by reference to Figures 5.6 and 5.8 — that turbulence is a very effective mixer of momentum, far more so than viscous mixing. The same applies to heat. Thus, the mixing by turbulence, illustrated in Figure 5.8 for momentum, also applies to heat or temperature: very close to the wall, turbulence is strongly damped by the wall, and mixing is dominated by the viscosity or conductivity; away from the wall, however, turbulence is well developed, and mixing is dominated by the action of turbulent eddies. If, for the sake of argument, we were to specify a temperature of 0°C at the inner wall of a tube, and the fluid temperature of the fluid inside the tube were at a higher temperature, the shape of the temperature distribution in the flow would be very similar to that of the velocity, as is shown in Figure 6.21.

As in the case of the velocity, turbulence leads to a flattening of the profiles inside the tube and a substantial steepening of the gradient of temperature at the wall, and it is the gradient at the wall that dictates the rate of heat transferred from the fluid to the wall, or vice versa. So, despite

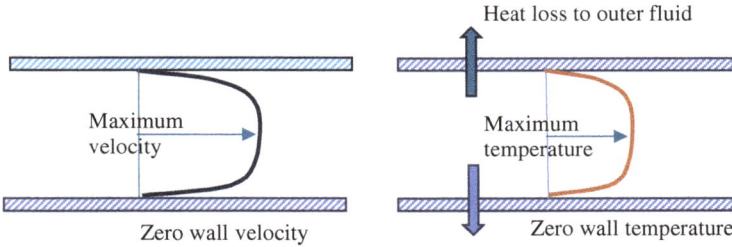

Figure 6.21.   The correspondence between the velocity and temperature profiles in a pipe in which the wall temperature is fixed at 0°C (assumed to be the datum value), corresponding to the wall velocity being zero.

the fact that the transfer of heat at the wall itself is by conduction only, the rate of transfer is indirectly enhanced by the turbulence away from the wall.

In practice, heat-transfer conditions are much more complicated than those shown in Figure 6.21. Even in a simple pipe, the complex reality may be that illustrated by Figure 6.22. We may know the outside temperature, far from the pipe. What we do not know, however, is what happens as the heat flows from the outside into the pipe, or vice versa. This is because we do not have information on the thermal resistance of the fluid close to the inner and outside wall. Therefore, we do not know the temperature profiles inside the pipe, in the outer layer and in the pipe wall.

In the circumstances shown in Figure 6.22, the heat transfer and thus the temperature development along the pipe are difficult to determine with any level of accuracy. If full information is sought on the thermal field, this can only be obtained by computational means of the type outlined in Chapter 3. A more practical approach is, however, to use empirical models that relate the heat transfer to the difference between the outside temperature and the average temperature in the pipe via a "heat-transfer coefficient". The crux of this approach is how to prescribe this coefficient with sufficient accuracy and in a manner compatible with the particular flow conditions in question. To this end, empirical relationships are called upon, which essentially relate the coefficient to the average velocity in the pipe, the flow conditions outside the pipe, the pipe diameter, the pipe-wall thickness and the material properties of the fluids and the solid — e.g., viscosity and conductivity.

Increasing the intensity of turbulence in a heating or cooling stream, by whatever means, improves the effectiveness of the heating or cooling.

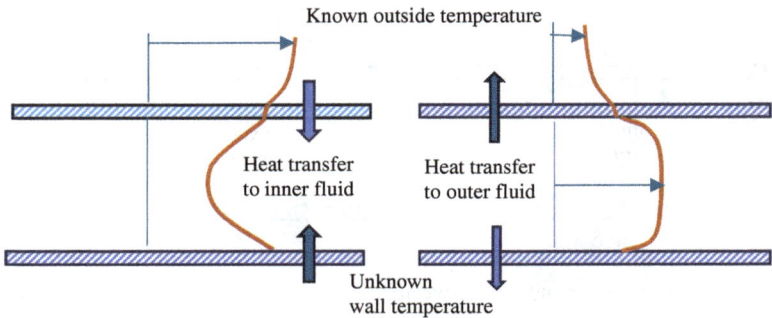

Figure 6.22. Two possible heat-transfer scenarios that are more general and more realistic than those shown in Figure 6.21.

This can be achieved by forcing the stream over surface-mounted ribs, fences or any other turbulators. In many cooling applications, the device to be cooled presents turbulating geometric features by default, as a part of its design or manufacture. Examples are a desktop computer and a laptop, in both of which the electronics and various components used to fix the microchips enhance turbulence by separation and recirculation of the type shown in Figures 4.32, 4.34 and 4.35. Also, computers almost invariably include fans that blow turbulent external air across the components to be cooled, in which case turbulence is enhanced by the blowing process, the shear it generates and the interaction of the air with the electronic components, including the convective transport of the heated air away from the electronic components and its replacement by cooler air.

Heat exchangers take a huge range of forms, but most consist, essentially, of pipes with straight and curved sections in a circular container. Some include elaborate fin systems attached to the tubes and designed to increase the area through which the heat is exchanged. The pipes carry fluid desired to gain or lose heat, while the other fluid flows outside along or normal to the pipes. In most circumstances, the flow in the pipes is turbulent. If the outer flow is normal to the pipes, it snakes between, and swirls across, the tubes and exchanges heat with the flow inside the tubes, as is shown in Figure 6.23. In this way, one of the streams is heated, while the other is cooled.

Air-conditioning units, refrigerators, dehumidifiers and heat pumps contain two heat exchanges: one for heating and another for cooling air streams pushed or pulled through the heat exchangers by fans. Figure 6.24

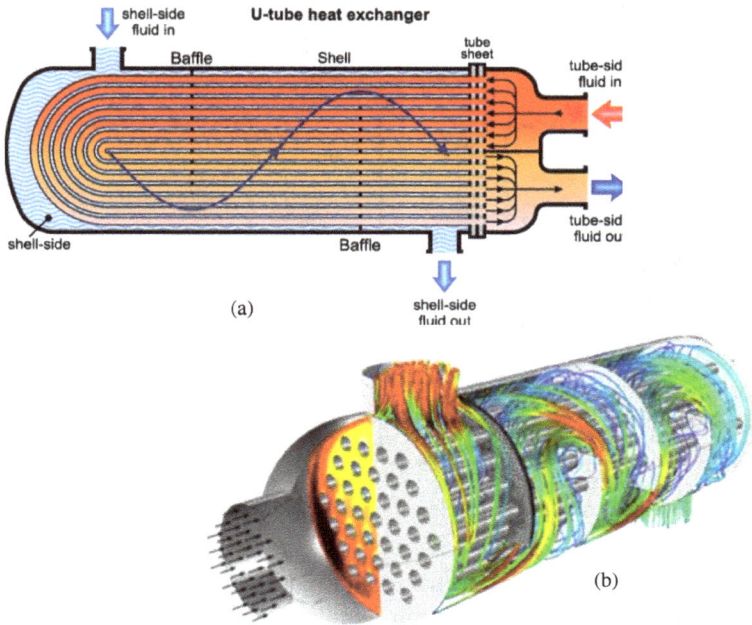

Figure 6.23.   Heat exchange between a bank of tubes carrying one fluid and another fluid flowing across the tube bank: (a) principle of operation; (b) computational simulation of the flow across the tubes.

Source: (a) Shutterstock (with permission); (b) image made using the COMSOL Multiphysics® software and provided courtesy of COMSOL.

shows the principles of how an air conditioner or refrigerator works. The working fluid is a refrigerant (e.g., a hydrofluorocarbon) that is made to change between a liquid and a vapour state by a compressor and expansion valve, respectively.

Following compression, the hot liquid passes through an external heat exchanger (coloured red), which transfers heat from the liquid to the outside ambient air. The cooled liquid then passes through an expansion valve (essentially, a small aperture) in which the liquid evaporates and from which the vapour exits at a much lower temperature. This temperature reduction is similar to that observed in a spray can when its content is discharged, resulting in the can becoming appreciably colder, alongside the cold spray. The cold refrigerant vapour then passes through the second heat exchanger (coloured blue) which transfers heat from the room or

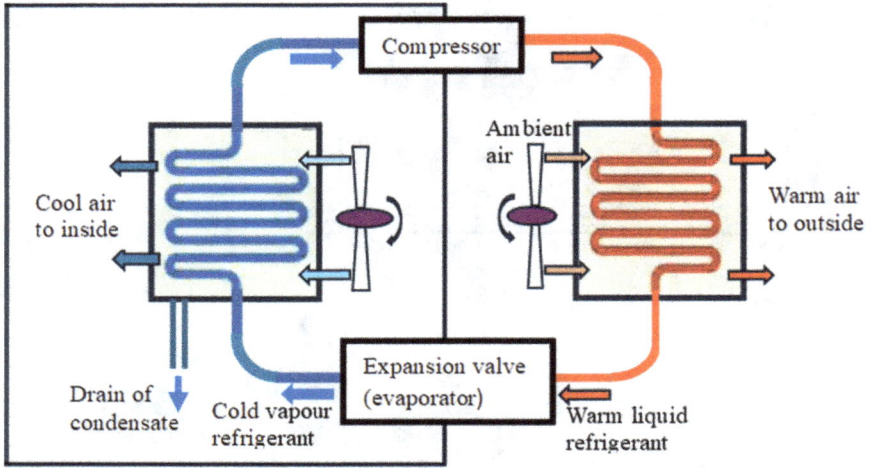

Figure 6.24.   Operational principles of an air-conditioning or refrigerating unit.

enclosure to the refrigerant, thus reducing the room temperature. The cycle is closed by the warmer vapour being compressed and transferred to its liquid state, a process that results in a strong increase in temperature.

In a heat pump, the outside and inside are reversed — i.e., the red heat exchanger in Figure 6.24 is in the enclosure to be heated, while the blue heat exchanger "cools" the outside (atmospheric) air by transferring heat from the outside air to the refrigerant.

A nuclear reactor is probably the most complex heat-exchanging system in existence. While the principles of its operation are relatively simple, the actual plant is extremely complex, partly because of the extremely stringent safety requirements that have to be met and also because its power output is very high, typically 1 GW. The working fluid in a reactor, normally water but in some liquid sodium, is heated by the fuel rods and flows in a closed primary (radioactive) circuit, as shown in Figure 6.25. The hot liquid then passes through a heat exchanger in which the heat is transferred to a secondary circuit. Cool water is pumped into this circuit, and this is converted into steam that is channelled to a steam turbine. The steam exiting the turbine is then condensed in a third heat exchanger, or a cooling tower, cooled with river or sea water, and the cooled water is finally pumped into the secondary circuit that is shown in Figure 6.24.

The nuclear reactor core is the most safety-critical heat-transfer component of the system, and its construction is extremely elaborate. This is

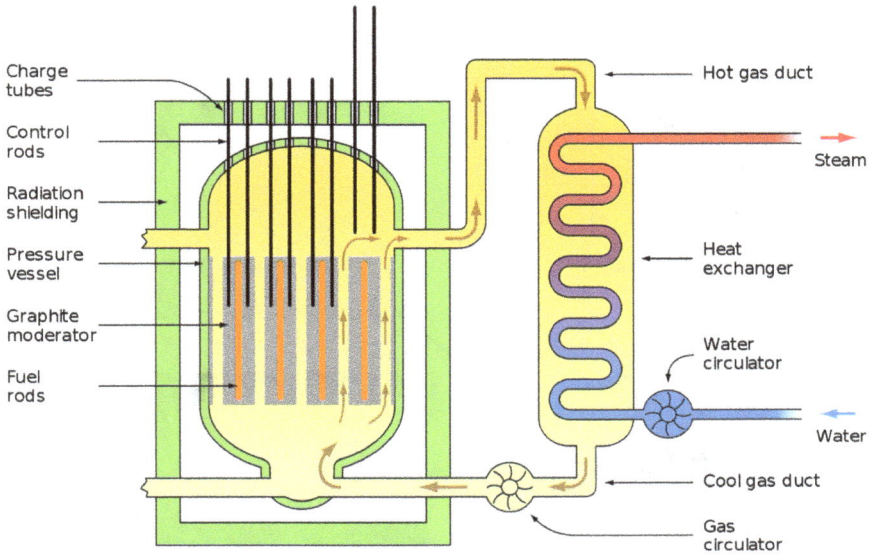

Figure 6.25.   Heat transfer in the primary circuit of a Magnox nuclear reactor core and in the secondary circuit of the heat exchanger generating steam for powering a turbine.

Source: Emoscopes Wikimedia CC-BY.

conveyed, albeit inadequately, in Figure 6.26(a), a cut through a reactor core. The core contains up to 200 heat-releasing rods of uranium or plutonium fuel and hundreds of neutron-absorbing control rods (e.g., made of boron). A critically important requirement of the core's operation is that the fuel rods are continuously cooled without hot spots being caused by local steam cavities that might cause the rods to distort, thus obstructing the cooling passages. The rods are therefore elaborately mounted within a cage comprising a matrix of passages within which the fuel rods are fixed.

As it is virtually impossible to conduct any experiments in a nuclear core, the design and operation of, and safety studies in, reactors rely on simulations using highly elaborate computer codes in which the fluid flow, heat transfer, multi-phase properties and neutronics are coupled. Two results derived from such simulations are included in Figure 6.26: insert (b) showing predicted temperature contours arising from heat being released by the fuel rods and being transported by the cooling liquid, and (c) showing the flow of cooling water and, in colour, the volume fraction of water and steam in the rod cluster.

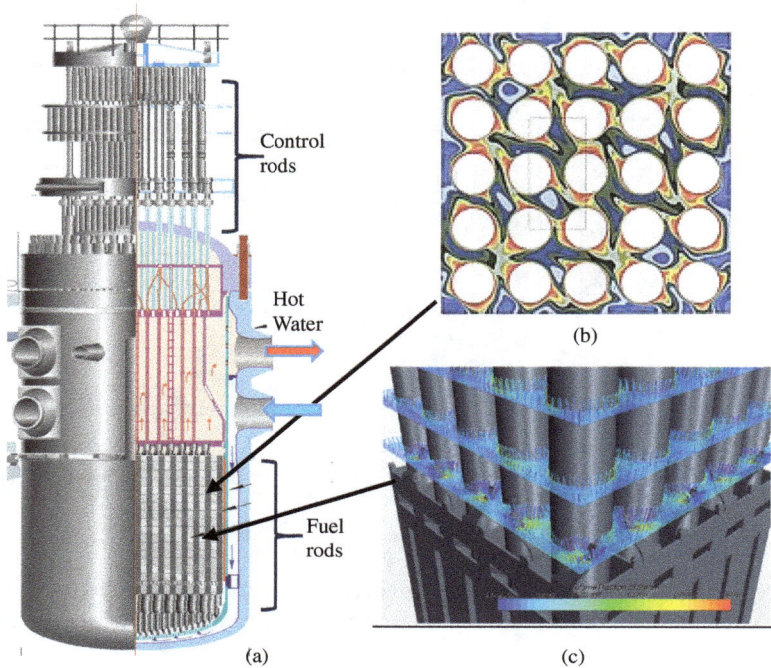

Figure 6.26. Nuclear power plant: (a) the core with control rods; (b) computational prediction of the temperature field around fuel rods; (c) computational prediction of the velocity and steam volume fraction in fuel-rod assembly.

Source: Wikipedia CC BY; (c) Siemens Industries Software Inc. (with permission).

6.7 Cast Your Eyes on Casting

Looking around you, are you able to spot solids that had transitioned through a liquid phase en route to reaching their final form? There are many: all metallic items, glass, plastics, concrete, cement, paints and paper. The fluid timespan of the life of some is unremarkable, arguably too mundane to examine in detail from a fluid-mechanic perspective. Cement and paints undergo simple mixing of solid or liquid additives with a solvent or water prior to their end state. Glass is a very viscous amalgam of sand, soda ash and limestone, all melted and mixed at a temperature of around 1700°C. The production of plastics is a more complicated multi-stage process, especially with respect to the chemistry involved, starting with the extraction of long-chain hydrocarbons from oil and coal, then cracking them into monomers, followed by polymerisation into pellets and finally melting and mixing the pellets with additives. Metals are produced by mixing ore, scrap metal and additives — e.g. carbon into iron — and smelting the mix in a blast furnace.

More interesting, from a fluid-mechanic standpoint, is the end-stage processing of some of the raw products. Plastics and glass are processed in a soft state that stretches the term "liquid". In most circumstances, the state of the material is like hot plasticine, and it is pressed into the desired shape in moulds, possibly in combination with blowing pressurised air into cavities. For example, Figure 6.27 shows a two-stage process of the production of a bottle. The moulding process creates a cavity in a malleable cylindrical segment, and this is followed by blowing pressurised air into the cavity within a second mould. While the physics of plastic deformation is challenging, this deformation is outside the realm of fluid mechanics, as interpreted in this book.

Of greater relevance to the present considerations is the casting of metals, which fully deserve the designation "liquid" prior to their final solidification. In fact, the viscosity of molten metals, although temperature and composition-dependent, is close to that of water, the value for molten aluminium being almost the same, while that for molten iron is about 5–10 times higher — still significantly lower than the value light oil. It follows that all of the flow-physical processes we have considered in earlier chapters — convection, turbulence, separation, recirculation, etc. — apply to liquid metals as well, and all these are pertinent when considering the behaviour of metals while being cast.

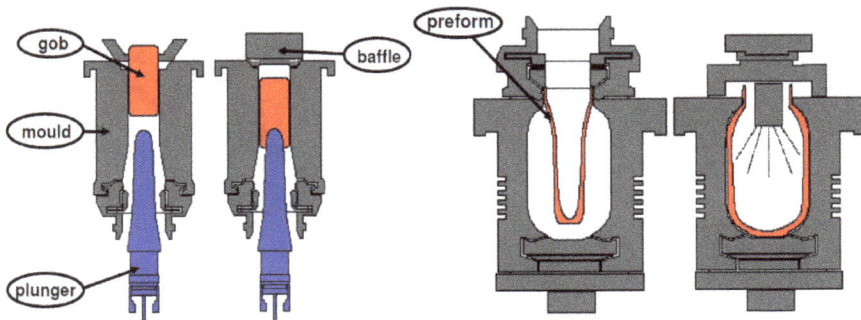

Figure 6.27. Two-stage "press-blow" process of producing a glass bottle in two moulds. *Source*: Groot [32] (with permission).

There are at least 10 distinctly different types of casting, the simplest and oldest (as old as 3000 BC) being sand-casting and the most elaborate — and modern — being high-pressure-die and vacuum-die casting, used predominantly to produce complicated aluminium-alloy components, especially for the automotive and aerospace industries. Due to the geometric complexity of the components being cast and the optical as well as physical inaccessibility of the moulds, the predominant tool for studying the fluid flow process in casting is by means of computational simulations, along the principles outlined in Chapter 3.

Figure 6.28 shows the principle of pressure-die casting and the outcome of a computational simulation for a geometrically complex gearbox housing, specifically the surface temperature during the solidification stage which is indicated by colour contours. Computational modelling is also applied to the flow in blast furnaces to gain insight into the processes by which the liquid steel, solids additives and gas within the liquid are being mixed. Some water-flow experiments in transparent mould models can be informative, but these are costly and rarely fully realistic because of the absence of heat transfer and solidification.

A major challenge in casting is to shape the mould and inject the metal in such a way that the mould is filled before the metal solidifies.

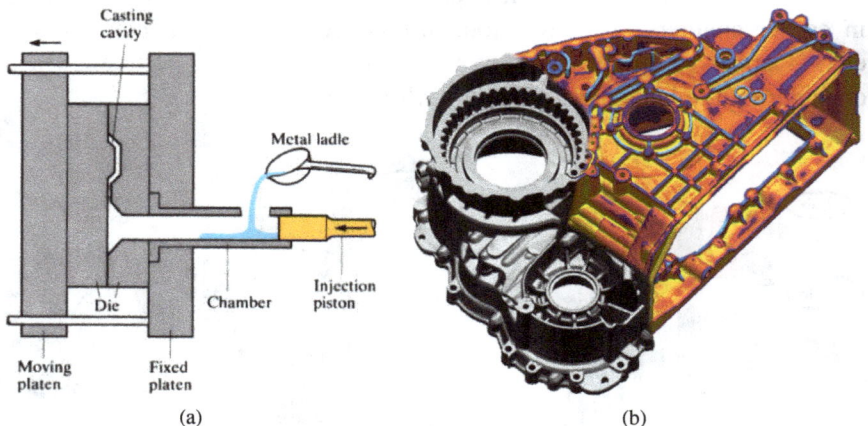

(a)   (b)

Figure 6.28.   Pressure-die casting: (a) schematic of process; (b) numerical simulation of casting an automotive gearbox housing in aluminium, with colours indicating temperature during solidification.

Source: MAGMA GmbH (with permission).

This requires elaborate optimisation studies that include thermal management of the mould-metal heat transfer, the speed of injection, the temperature of the liquid metal and the provision of runners (ducts) in the mould at locations that are remote from the injection point, so as to eliminate air pockets and allow the metal to flow to all parts of the mould.

As the metal flows into the mould, it tends to fragment at the air/metal interface. If the process is poorly configured, the mould fails to fill before solidification occurs. This is illustrated by the computational simulations, as shown in Figures 6.29 and 6.30. The avoidance of this

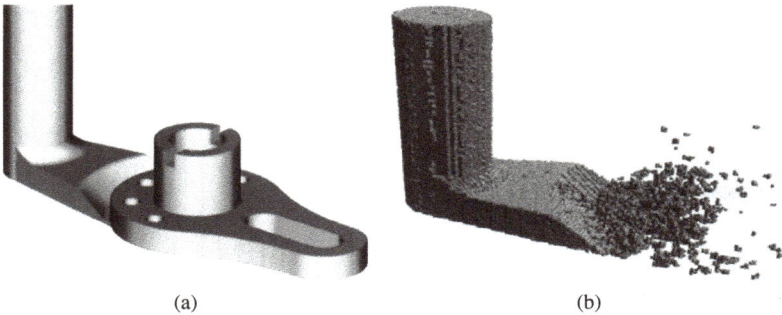

(a)   (b)

Figure 6.29.   Pressure-die casting: (a) desired component; (b) computer simulation of defective ("short-shot") injection process.

Source: Cleary *et al.* [33] (with permission).

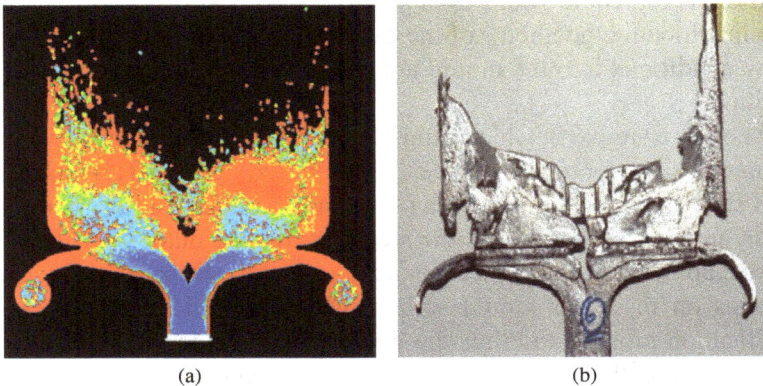

(a)   (b)

Figure 6.30.   Computer simulation of defective ("short shot") pressure-die casting and comparison with the real (experimentally derived) component at the same injection conditions.

Source: Cleary *et al.* [34] (with permission).

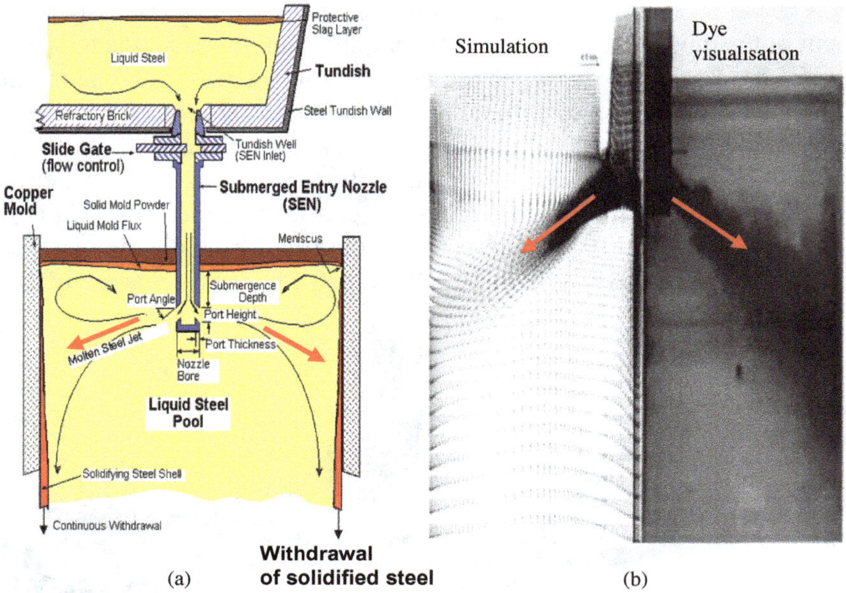

Figure 6.31. Continuous steel casting: (a) schematic of process and qualitative flow features due to injection from a central pipe nozzle into a mould; (b) computational simulation of the nozzle flow compared to measurements in a water model.

Source: Thomas [35] (with permission).

"short-shot" outcome requires many repetitions of the computational cycle in which the geometry of the overflow runners is modified and the inflow conditions into the mould are examined over a range of injection conditions.

Another example in which fluid-flow mechanisms are important is the continuous casting process of steel, as shown in Figure 6.31(a). Here, the liquid steel flows freely through a nozzle into a copper mould, which creates an annular jet that directs the steel towards the outer wall. The ring-shaped gap at the bottom of the nozzle gives rise to an annular jet that impinges on the outer mould wall at which solidification occurs. The solidification process and thus the downward-directed withdrawal of the solid depend on the flow features induced by the nozzle. As in the case of pressure-die casting, computational simulations are used, here too, to examine the consequences of modifying the geometric parameters of the nozzle. Figure 6.31(b) shows a comparison between a simulation and measurements in a water model for one particular nozzle — essentially, a validation of the ability of the simulation process to represent the principal flow dynamics that might be expected in reality.

6.8 Bobbing Up and Down: Waves

Few would disagree with the observation that there is something magically soothing about large expanses of blue water in lakes and seas. Yes, there is the mysteriously calming quality in the stillness of the water surface, but more often than not, it is the dynamics of waves, especially the awesome sound and fury of large breaking waves crashing onto the shore, shown in Figure 6.32, that offers the most enticing and fascinating characteristics of the sharp interface between the water surface and the atmosphere above it. Unsurprisingly, therefore, masses of holidaymakers (and surfers) flock to seaside resorts during the summer, and real-estate prices rise sharply with the proximity to the shores of lakes, seas and oceans. These enjoyment aspects aside, the sea and ocean surface is critically important to the climate, to the formation of rain clouds by water evaporating from the surface and to the exchange of O_2 and CO_2 across the interface — waves being especially influential to this gas exchange.

Figure 6.32.   A breaking wave on a slopping shore.
Source: iStock (with permission).

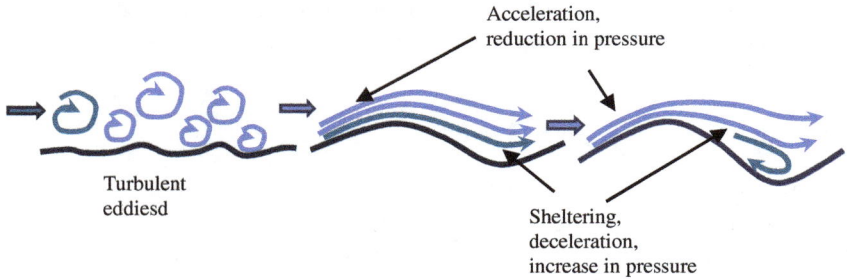

Figure 6.33.   The mechanisms by which wind initiates and amplifies waves.

With the exception of tsunamis and solid bodies protruding through the water surface or moving on it, both considered later, waves are formed by the action of wind on the surface. While many of us have an intuitive understanding of the formation and growth of waves, the physical mechanisms are complex and not fully understood but involve two principal ingredients: shear and friction on the water surface and pressure differences within the wind as it flows from crests to troughs and vice versa.

As shown in Figure 6.33, the process starts with the formation of surface ripples caused by turbulent eddies in the shear layer above the water surface. As the waves grow, the acceleration on the windward side of the wave causes a suction force on the surface, with the maximum suction acting on the crest — much like that on the suction side of a wing or over a hill (see Figure 4.14, Section 4.3, and Figure 4.21, Section 4.4.2). The acceleration also causes the shear force of the windward side to increase. As the wind progresses into the sheltered leeward side towards the trough, the flow decelerates, the pressure recovers (i.e., increases) and the shear force declines, possibly reversing its direction if the airflow separates from the crest. The combination of shear force and pressure differences over the moving waves progressively feeds energy into the waves, at the expense of energy loss from the wind, and the waves grow in amplitude, wavelength and wave speed. Eventually, in idealised conditions, a state is reached in which the speed of the increasingly long waves approaches the wind speed, and there is a "resonance" between the two in which the transfer of energy into the waves is predominantly by wind shear on the surface.

As an aside, it is interesting, and of relevance to the present topic, to remark here that sand dunes, the regular ripples on their windward

Figure 6.34.   Sand dunes and wavy ripples driven by wind shear.
Source: Wikimedia Commons, CC-BY.

slope — shown in Figure 6.34 — and the wavy pattern on a sandy river-bed or beach subjected to flowing water are formed by processes that share some elements with the mechanisms shown in Figure 6.33 and described above. However, it is also important to recognise the limits of this commonality. Although sand can behave like a fluid, especially when infused with air or water, it is not a water-like liquid in terms of its physical properties. Sand does not move, unless subjected to considerable shear, does not propagate in waves when the shear is removed and is only weakly affected by wind- or water-induced pressure gradients. The dominant mechanism is sand being dragged up the windward slope by surface shear and then flowing into the sheltered low-shear leeward trough where is deposited onto the surface. Because of the very different physical properties of water and sand, the length scales (e.g., wavelength) of sand dunes and sand ripples cannot be analysed in terms applied to water waves. Hence, caution is called for when juxtaposing water waves with sand "waves".

The fact that water waves contain energy is an important one to appreciate. It takes energy to create waves, be it by wind or by moving a solid in the water, and this energy propagates with the waves over long distances. Waves are damped only slowly, by friction within them, but this friction is low, especially in long waves over deep water, because the speed of, and thus shear within, the fluid below the waves is low relative to the speed of the wave for reasons explained in the following description.

One particular case of a solid creating waves is a ship moving at moderate speed. This is also a very important case, in terms of wave energetics, because shipping contributes roughly the same amount of CO_2 emission as air transportation does (around 3% each). The drag on a ship consists mainly of two components: friction drag and wave drag, both requiring the expenditure of energy, because the power expended is the product of the drag force and the speed of the ship. The relative magnitude of the two drag components depends greatly on the speed of the ship, the smoothness of the hull and the length and shape of the ship's nose. In particular, a bulbous bow (mostly below the surface) tends to reduce the wave drag. The wave drag can easily exceed 50% of the total drag at high speed, and this exemplifies well the relevance of wave energy to fuel consumption and thus environmental issues.

A sea-surface wave is, in general, a complex three-dimensional amalgam of many interacting waves. In fact, the oblique collision between two large, high-amplitude, waves approaching each other at a particular angle, is the presumed cause of extreme-height ("rogue") waves that are feared by mariners of even the largest ships sailing the oceans. However, there are many circumstances in which the predominant behaviour is characterised by long, wind-induced, periodic waves that propagate in a single direction. Such waves justify some key characteristics to be explained by reference to an idealised two-dimensional model, and this is done next.

An intriguing aspect of wave motion is that the speed of propagation of waves involved very little mass being transported with the wave — unless the focus is on the final stage of a shallow wave breaking on the shore. Figure 6.35 shows three time snapshots of a wave propagating to the right in deep water ("deep" being defined later).

Imagine a ping-pong ball — the red dot — placed on the water surface. As the wave propagates, the ball moves upwards and forward as the wave grows towards its crest but then moves backwards behind the crest. In effect, we are faced with two wave motions: one vertical and the other horizontal. The result is a spiralling, nearly closed orbital motion in which the ball, and thus the mass on the surface and below, drift forward much more slowly than the speed of the wave. Anyone observing a duck or a gull floating on a wavy water surface will observe that the bird's motion is characterised by up-and-down "bobbing" motions on the wave, with

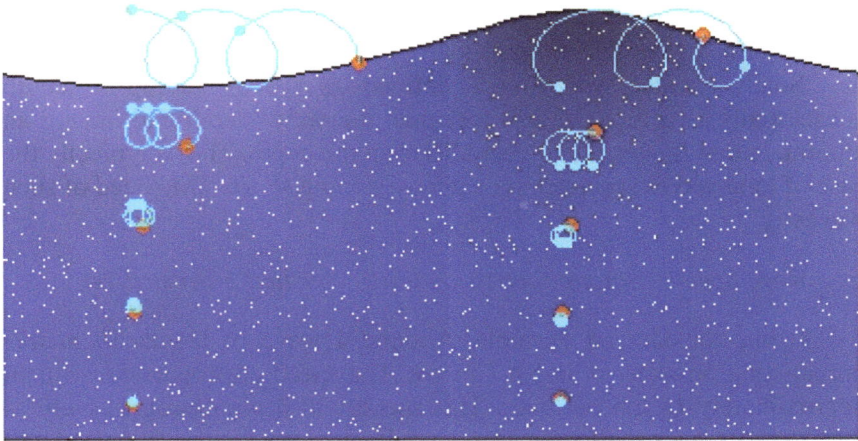

Figure 6.35.   The motion of water particles on and under a propagating wave in deep water. The image shows the trajectory of virtual markers after the wave has moved forward by, roughly, three wavelengths.

Source: Images derived from an animation on Wikimedia Commons, based on Fenton [36].

any lateral drift being insignificant. As the depth increases, the water particles likewise move in nearly closed orbits but with decreasing radius so that any mass movement under the wave is very small. Thus, fish swimming at a depth of more than one-half of the wavelength will experience very little movement due to the travelling wave. However, this does not apply in shallow water in which the sub-surface motion is more intense.

Another intriguing aspect of wave motion is that the speed with which a wave propagates depends greatly on the wavelength relative to the depth of the water body. The behaviour shown in Figure 6.35 is predicated on the water body being "deep". A deep body is one in which the depth is at least of the same order as the wavelength. In this case, the speed of the wave is given (approximately) by

wave speed = square root {gravitation constant × wave length/(2 × π)}

where the gravitational constant (acceleration) is, in metric units, 9.81 metres/second square and π is 3.14.

Consider the example of a wave of length 10 metres. The speed of wave propagation is then around 4 metres per second or 14 km per hour. Correspondingly, a wave of 100 metres in length will propagate at a speed of 45 km per hour. The associated mass drift, on the other hand, will only be a tiny fraction of this value.

"Shallow-water waves" are defined as waves having a length substantially larger (at least twice) than the water depth and having an amplitude that is substantially smaller than the depth. This is typically the case of long waves approaching a gently sloping shore. The speed of such waves is

wave Speed = square root {gravitation constant × water depth}

Again, consider an example. Assume a wave of length 10 metres propagating on water of depth 1 metre. The speed of propagation is then approximately 11 km per hour. This speed will decline progressively as the shore becomes shallower. Thus, at a depth of 0.5 metres, the speed will decline to 8 km per hour.

The above considerations on wave speed serve the purpose of explaining why waves increase in height, become steep and then break in shallow water. As shown in Figure 6.36, a long wave approaching a sloping shore slows down, and its wavelength is thus decreasing. Faster waves approaching from the left "pile up" on the slower waves. As the total energy of the decelerating waves is broadly maintained, their height (i.e., their potential energy) has to increase. In the final stage, bottom friction progressively slows down the water close to the bottom relative to the water above it, and the crest close to the surface races forward with air replacing the receding near-bottom water, thus causing the wave to break.

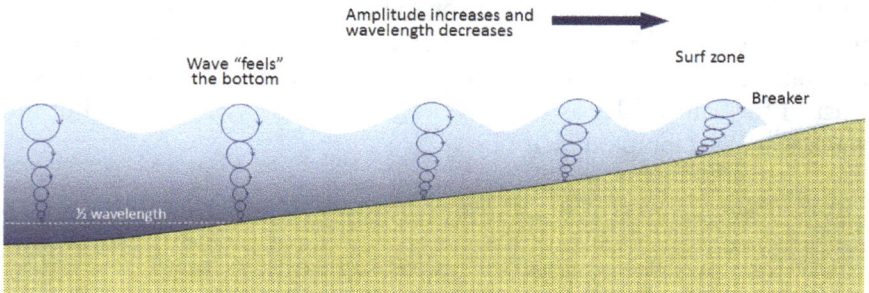

Figure 6.36. The behaviour of a wave in the transition between deep water and shallow water, including breaking.

Source: Earle [37], CC-4.0.

The breaking process is characterised by intense, noisy turbulence, which dissipates the wave energy into heat — the final act of the life of the wave. The breaking process is further enhanced by water on the shore left behind from a previous breaking wave flowing back down the slope as a thin film, which thus further decelerates the water close to the bottom in the oncoming wave about to break.

One particular — and particularly destructive — type of wave is a tsunami (literally translating as "harbour wave" from Japanese). Such a wave is usually generated by a rapid, large-scale vertical displacement of a part of the seabed as a consequence of continental subduction or an explosive volcanic eruption in the ocean (e.g., the Krakatoa explosion in 1883). The upward displacement of mass over a very large area generates a powerful wave which is deceptively gentle in deep water, having a low amplitude of less than 1 metre, but having a very large wavelength, typically 100–300 km, relative to around 100 metres for wind-induced waves. Due to its very large wavelength, a tsunami needs to be viewed as a shallow wave. Thus, for example, assuming a water depth of 10 km, the speed of a Tsunami might be 1000 km/hour, i.e., the speed of a passenger jetliner! The huge amount of energy contained in the wave, its phenomenally high speed and the rapid slow down it experiences as it approaches the shore result in frightening levels of destruction, as we know especially well from recent earthquake events in Tohoku (Fukushima) in 2011 and the Indian Ocean (Sumatra) in 2004. Thus, as is often the case with natural fluid-mechanic phenomena, there can be a painful price to pay for their benefits. It has frequently been observed that the first phase — the harbinger — of a tsunami reaching land is the drainage of water and retreating of the water edge over large portions of the shoreline, especially when it is sloping gently. This occurs in about half of all tsunamis in which case the trough of the first tsunami wave leads to the crest. The result of this reverse streaming is that the approaching crest is magnified, leading to a more powerful breaking of the wave and thus worsening its destructive effects.

A final interesting topic under the heading "waves" is concerned with the manner in which waves propagate in a moving stream or, equivalently, when a solid moves in a stationary water body. These considerations then allow us to interpret a number of intriguing water-surface phenomena, observed frequently when fast-flowing water is opposed by solid obstacles or by an elevated downstream water height.

Imagine that you are perturbing a wide stream of shallow water of constant depth by poking the surface periodically with a pencil or any

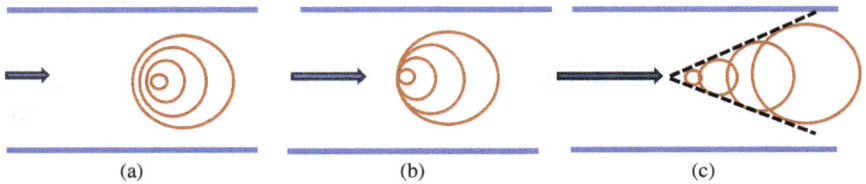

Figure 6.37.   Propagation of weak surface waves resulting from periodic perturbations in a stream moving, (a) at a speed lower than, (b) identical to and (c) higher than the wave speed.

other sharp object. Viewed from the top, there are three distinctly different wave patterns possible, depending on the relative speed of the water and the wave-propagation speed, as shown in Figure 6.37(a)–(c). The left sketch (a) shows the behaviour when the wave speed is higher than the speed of the stream, in which case the waves spread in all directions, including against the flow. The centres of consecutive waves are displaced to the right because the waves are translated to the right by the water motion as the waves spread. When the speed of the water is the same as the wave speed, as is shown in sketch (b), the waves just stop propagating upstream. Finally, sketch (c) shows the pattern that arises when the water speed exceeds the wave speed. These scenarios are equivalent, respectively, to the propagation of weak sound waves in subsonic, sonic and supersonic flow, the dashed lines in sketch (c) being equivalent to the "Mach cone". Consistently, the three states are thus referred to as "sub-critical", "critical" and "supercritical".

We know that placing an obstacle in a supersonic flow creates a shock wave, as shown in Figure 6.38(a), which is a visualisation of an experiment on a space-vehicle model re-entering the atmosphere. A shock wave is a strong discontinuity that reflects a pile-up of pressure (sound) waves that are generated at the body, propagate upstream in the slow subsonic flow close to the body, but are then unable to propagate further upstream, because of their insufficient speed, into the supersonic flow, thus creating a physical incompatibility. A similar picture presents itself when a super-critical flow approaches a bridge pier — or any other obstacle, such as the rounded bow of a ship moving at high speed through stagnant sea water — as is shown in the upper-right image of Figure 6.38(b), a view from above onto the water surface. Here, the water waves are unable to propagate into the supercritical flow, and a discontinuity called "hydraulic

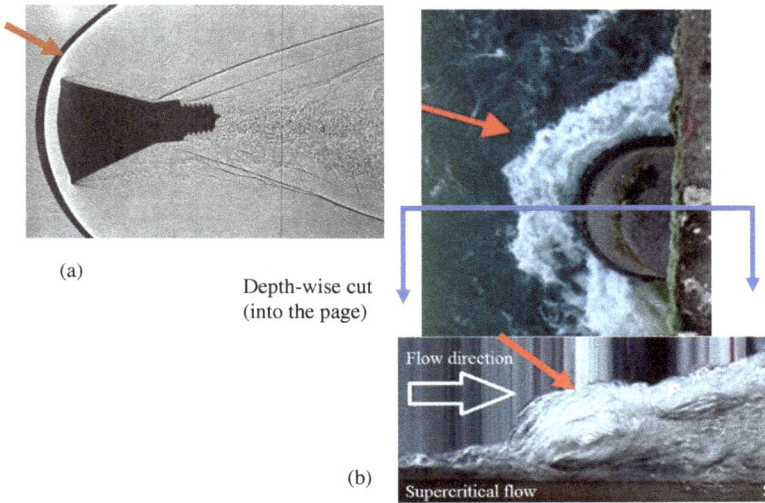

(a)

Depth-wise cut
(into the page)

(b)

Figure 6.38. Illustration of the equivalence between (a) a shock in supersonic flow and (b) a hydraulic jump in a super-critical water flow, in both cases caused by an obstacle.

Source: (a) NASA; (b) Link *et al.* [38] (with permission); Schulz *et al.* [39] (with permission).

jump" is created. Such jumps are readily observed upon placing a stone in a fast-moving shallow stream and even when a finger is inserted into the thin layer of water flowing radially outwards on the surface of a flat kitchen sink after the kitchen tap is turned fully on. If a vertical cut is taken through the jump, as is shown in the lower image of Figure 6.38(b), the jump reveals itself as a near discontinuity separating a thick layer of slow-moving (subcritical) water on the right from a fast-moving (super-critical) thinner layer on the left. The difference in surface elevation is equivalent to the pressure jump across the shock wave in the supersonic (or, rather, transonic) flow. This process is easily observed in a kitchen sink when a fast stream of water is made to impinge on the flat bottom of the sink, as shown in Figure 6.39. This causes a fast — supercritical — thin film to flow radially outwards on the bottom. Due to the constraining influence of the vertical sink sides, combined with friction on the bottom and a deceleration in the radially expanding film, the flow approaching the sides is forced into a subcritical state, and this has to occur across a discontinuity — a hydraulic jump very similar to that shown in Figure 6.38(b).

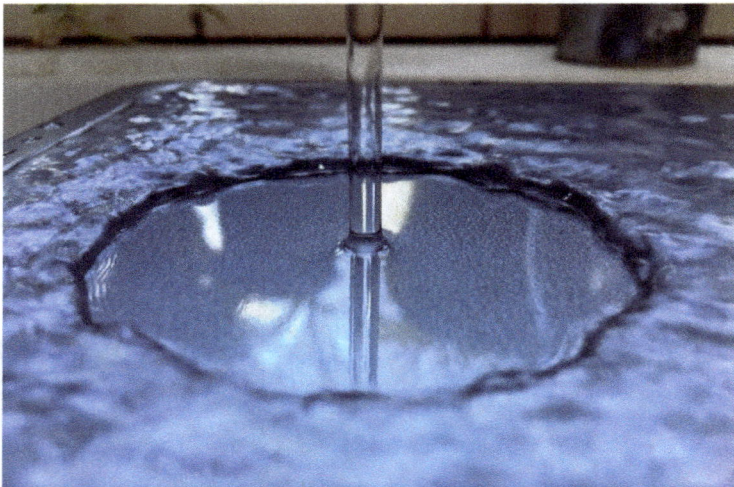

Figure 6.39. A hydraulic jump realised in a kitchen sink.
Source: Protons for Breakfast, Wordpress.com (with permission).

Since the above water-based phenomena can be visualised so easily in very simple configurations, in contrast to high-speed air flows, the equivalence between sound waves and gravity waves is almost invariably exploited in lectures on transonic flow theory presented to undergraduate students.

A fascinating, and rare, manifestation of an interaction akin to a hydraulic jump is a "tidal river bore", shown in Figure 6.40 (the name "bore" originates from the Old Norse word "bara", meaning a wave or swell). This occurs in long, gently sloping, shallow stretches of natural rivers in which bed friction is relatively high and water flows, often gently, towards the sea against a high tide at the river mouth. Since the wave speed in the shallow water is low, relative to the speed of the penetrating tidal stream, and the shallow stream also flows against the tide, a shock-like step is formed at the front of the bore, the structure of which is similar to the flow in Figure 6.38. Two differences are, first, that the bore is supplied continuously and independently from the river water by the high-elevation tide and, second, that the low-elevation river water flows against the bore at approximately constant speed rather than slowing down because of the radial spread in the flow in Figure 6.39. Strong bores are beloved of a section of the surfer community because the front of the bore

Figure 6.40. A "bore" in a tidal river formed by the interaction of upstream-moving tidal water with downstream-moving river water.

Source: T. Gertridge, Wikimedia Commons CC-BY.

can be ridden in a manner similar to that presented by large ocean waves, though at a much lower risk to life and limb.

6.9 Wind in My Sail

Since we are discussing water and waves, we might just as well continue with a subject that combines the heady mix of water, wind, sport, science and even more fun than riding a bore. Lest anyone doubts that a subject almost as old as mankind itself has anything to do with advanced science, reference to some of the images and videos showing the hugely sophisticated America's Cup sailing yachts should suffice to quash any such doubts. Any one of these yachts is the product of years of scientific endeavour by teams of academic researchers and industrial engineers, involving complex fluid mechanics, computer codes, model experiments in wind tunnels and water flumes, and endless iterative optimisations and improvements. As is shown in Figure 6.41(a), the NZ entry in 2021 is a hybrid between a boat and a flying machine, capable of reaching speeds approaching 50 knots. It does not merely use the wind for propulsion, but it is actually riding on hydrofoils — underwater wings — and is thus "flying" above the water surface.

Prior to embarking on a description of the principles of how sailing boats exploit the wind, it is appropriate to point out that fluid mechanics contributes to many sports in many different ways. The performance of golf, cricket, tennis and footballs is one example already discussed in

(a)   (b)

Figure 6.41.   Two very different sailing boats: (a) New Zealand's high-tech boat entry to the America's Cup sailing competition; (b) a simple sailing boat with a mainsail and a headsail.

Source: Emirates Team New Zealand (with permission); Wikimedia Commons, CC-BY.

Section 5.4 (see Figure 5.7). Others are the aerodynamics of cycling — in particular, the cycle shape itself, the helmet, the optimal rider position and the gear worn by the rider — drag reduction on swimwear by use of micro grooves in the fabric (to which we return in Chapter 7), the hydrodynamic of competitive rowing boats, kayaks, powerboats and jet skis — and, of course, the aerodynamics of racing cars (see Chapter 3, Figure 3.4).

An exceptional aerodynamic feature of sailing is the extreme variability of the conditions under which the sailing boat has to manoeuvre. In general, there is one predetermined direction in which the sailor wishes to head. However, the wind may have other ideas: it may not simply be gusty and directionally viable, but its average direction may be opposite to the direction of travel — maybe not fully opposite but at an angle not far different from pointing backwards. This poses extreme challenges to both the sail and the operator.

While a floppy fabric sail is not a wing, the two share common characteristics: both deflect the oncoming flow, thus producing a force; both are curved, although the pressure and suction sides of a sail have (normally) the same shape and are separated only by a flimsy fabric layer. Sails vary greatly in shape, size and installation on different vessels. There are at least eight different types, and this excludes rigid or semi-rigid, high-tech sails that may have wing-like profiles, thus featuring true suction and pressure sides. The main types on smaller boats are the mainsail, mounted directly on the mast in the centre of the boat, the headsail mounted flexibly on ropes towards the bow, and the spinnaker, a large baggy sail mounted ahead of the mainsail on a boom hinged on and extending vertically from the mast. The last is used when wind is blowing in the direction of travel, and this operates, aerodynamically, much like a flat surface subjected to an oncoming flow. In the discussion to follow, we examine the simple, standard configuration shown in Figure 6.41(b).

When the wind is broadly in the direction of travel, sailing is straightforward: the sails (normally including the spinnaker) are facing the oncoming wind, the wind is losing momentum, and the loss of momentum is reacted to by a forward-facing force on the sails, driving the boat along. However, a more challenging condition, not easily understood, is one in which the wind is blowing in a direction opposing that of travel, a scenario shown in Figure 6.42. Again, the boat is here assumed to have a mainsail and a headsail only.

As is shown in Figure 6.42, there is a strong side force on the boat, so what prevents the boat from drifting sideways faster than moving forward (assuming it stays upright)? The resistance offered by the length of the body of the boat is one reason. But a more important reason is the presence of a wedge-shaped keel (and ballast at its bottom) protruding downwards from the bottom of the boat. On the largest sailing boats, a keel can

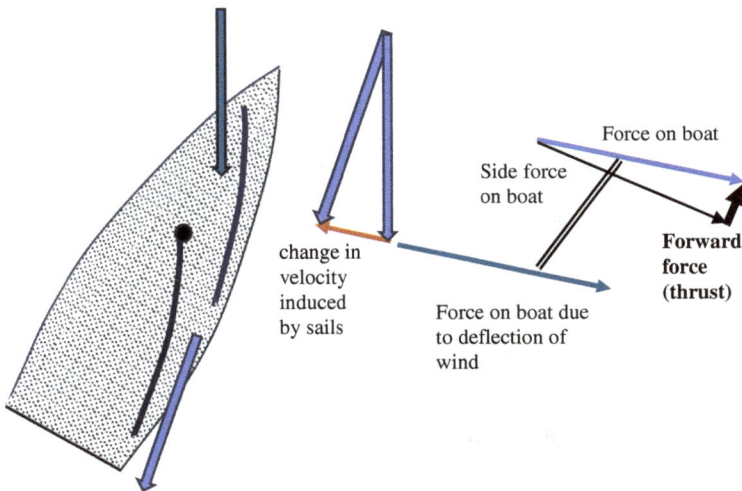

Figure 6.42. The aerodynamic conditions of a sailing boat of the type shown in Figure 6.41(b) when sailing against the wind.

weigh several tons and extend to a depth of 7 metres. Its function is to resist the sideways drift. Also, when the boat is tilting in the clockwise direction by the moment of the sideways force acting on the sail, the weight of the keel, combined with the buoyancy of the boat, create a moment (torque) in the anti-clockwise direction that opposes the clockwise tilting. This is why the boat remains upright — well, at least when it is under the control of a competent sailor.

6.10 Fresh Water: The Elixir of Life

Fresh water is arguably the most critical resource for sustaining plant and animal life on the Earth's land mass. Its availability is governed by a complex set of interactions between the oceans and the climate, referred to as "The Water Cycle", the main components of which are shown in Figure 6.43.

It should be readily recognised that these interactions are hugely variable, depending upon climatic conditions — rainfall and draught conditions, in particular — the geographic location and its topography, the

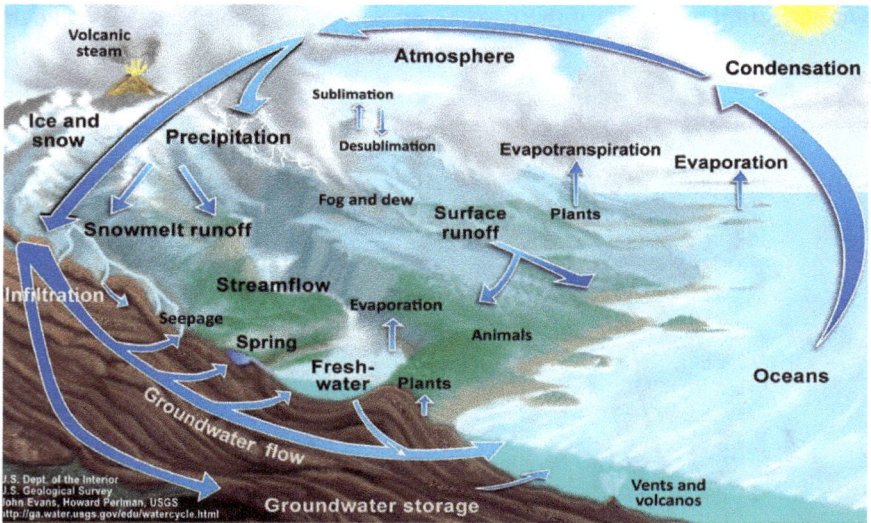

Figure 6.43. The water cycle: evaporation, condensation, precipitation, groundwater, surface runoff, snowmelt, underground and overground storage.

Source: NASA, US Geological Survey.

time of year, cyclical and volatile weather phenomena (e.g., monsoon, typhoons and hurricanes), the ratio between the surface areas of land and sea, the structure of the soil — its ability to absorb and retain water and the presence of underground streams and aquifers, in particular —— the river system and its regulation, the distribution of built-up urban areas, forests and agricultural activity. Alongside the natural processes identified in Figure 6.43, there are numerous man-made factors at play, including water storage in artificial reservoirs and tanks, river management, flood regulation and prevention by channelling and rerouting streams and rivers, tidal-power generation, domestic and industrial water demands and distribution, desalination and sewage treatment.

The science that describes the natural interactions shown in Figure 6.43, including data collection, statistical processing, modelling water movement and the description of the consequences of human interventions affecting the availability of water, is called "Hydrology". In contrast, the technological exploitation and management of water and the control of how water is distributed to users is called "Hydraulics". The later technology also includes hydro-power generation and the use of liquids in power machinery to lift and shift loads.

Due to the highly localised conditions, the huge number of variables and parameters affecting the availability and management of fresh water, it is virtually impossible to give a general description of hydrological or hydraulic scenarios, or even one that is representative of significant sub-areas of the science and technology involved. It is a reflection of the complexity of hydrology that there are at least two dozens of very diverse hydrological models which are used by the scientific community for similar purposes to analyse and predict the effects of the "input" parts of the water cycle on the "output" — the water that eventually flows into rivers and groundwater storage. Yet, the availability and distribution of freshwater are so central to our lives that they cannot remain unmentioned in this book.

Of the many individual branches of hydrology and hydraulics that one might discuss in greater detail, we consider here one particular subject that has an exceptionally strong impact on our daily life, namely, the distribution of water for domestic and commercial use by a pipe network — a branch of hydraulics.

It has to be noted first that no two water distribution networks are similar because no two towns or cities have anywhere near the same topography. Every village, town or city grows organically, and the water

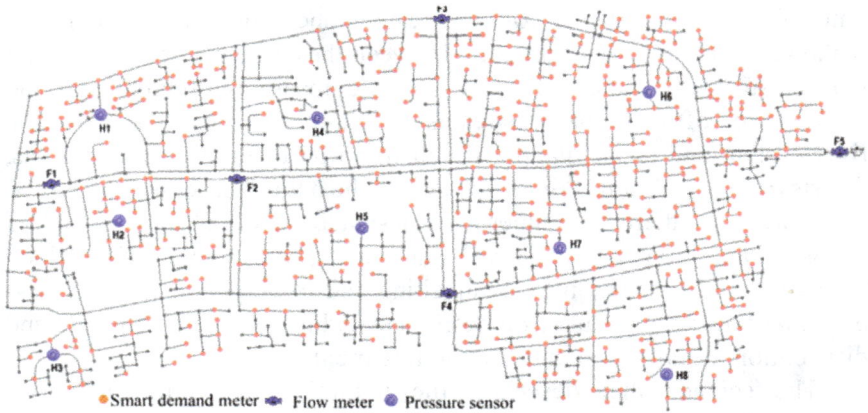

Figure 6.44. The main elements of a water distribution system in a Chinese town serving 107000 inhabitants.

Source: Zhang *et al.* [40] (with permission).

network has to grow and adapt with it, with more pipes, more branches, more pumps and more reservoirs added with time. In any town of a significant size, the distribution network will be very complex, comprising many pipes, loops, junctions, pumps, regulating valves and reservoirs. One can pick an almost infinite number of examples to illustrate this complexity.

One fairly simple network in a relatively small town in China is shown in Figure 6.44. This system has 1 reservoir, 1 pumping station, 1140 supply pipes, 8 pressure sensors and 525 smart demand metres. It supplies 23000 cubic metres per day to a population of 107000.

A fact to bring out in relation to any water distribution system of the type shown in Figure 6.44 is that the structure of the associated wastewater network looks, as one might expect, very similar to the former. Indeed, in practice, the two are necessarily designed, constructed and extended in parallel. This correspondence is conveyed, in a generic sense, by the illustration in Figure 6.45 in which a highly simplified, water network is shown together with the associated wastewater network.

An important operational requirement of a water distribution network is that the reservoir(s) and pumps are able to respond at all times to the highly variable water demand profile and that the pipes, especially the major arteries, are able to accommodate the large flow rates that are

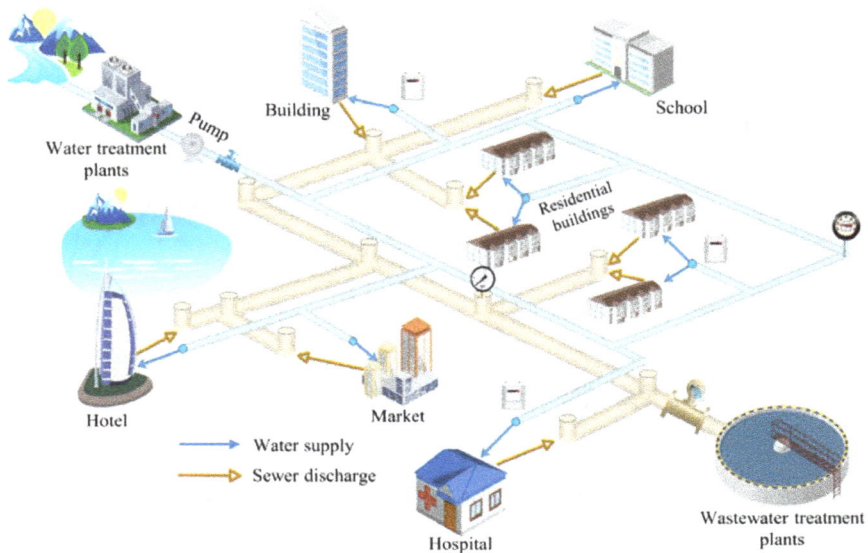

Figure 6.45. A schematic representation of the conformity of freshwater and wastewater networks in an urban environment.

Source: Zhang *et al.* [40] (with permission).

dictated by the usage of the many households connected to the network. Moreover, as the urbanized area expands, the network has to be extended in such a way that no other part of the existing network is compromised. This is no mean task, and the information needed to secure a reliable operation of the network relies on complex models that are able to quantify a large number of possible scenarios in which the many nodes and pipes carry and distribute water at different rates which change with time and spatial location.

A key ingredient of the analysis of a network is a relationship that links the flow through any pipe and the pressure difference between the end points that drive the flow. Thus, by reference to the circled branch in the simple model shown in Figure 6.46, there is a unique pressure difference ($Pressure_1 - Pressure_2$) for any given flow rate in the pipe. This pressure difference is due to the frictional resistance in the pipe and any valve resisting the flow in the pipe. The frictional resistance depends on the pipe diameter, the pipe length, the roughness inside the pipe and the characteristics of any flow-resisting device within the pipe — a

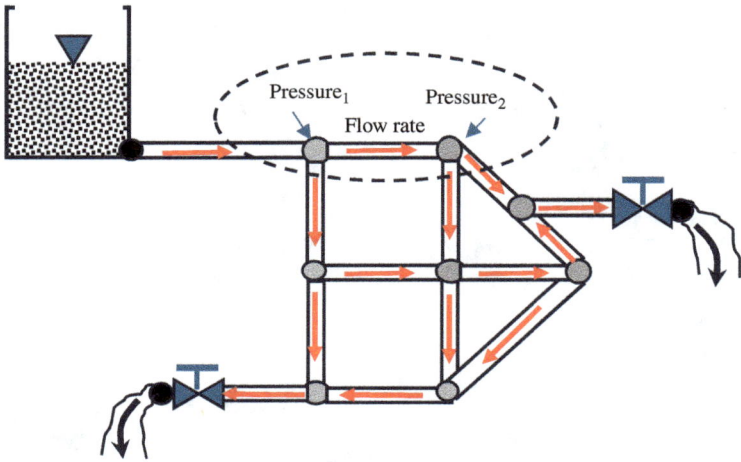

Figure 6.46. A simple model of a water distribution network used to illustrate how interconnected pipe systems are analyzed (see text).

dependence expressed via an empirical (e.g., tabulated) "friction factor". For a plain pipe, the relationship is of the form

$$\textit{pressure difference}_{1\rightarrow 2} = (\textit{constant}) \times (\textit{friction factor}) \times (\textit{flow rate})^2$$

where the *constant* contains the known water density, pipe diameter and pipe length.

In Figure 6.46, there are 14 pipes and 11 pressure nodes. Two of the nodes (in black) are connected to partially open water valves (taps) at the exit of which the pressure is known to be atmospheric, while the pressure at the reservoir node (in black) is fixed by the height of the water column in the reservoir, this pressure difference essentially dictating the flow through the network. There are, therefore, 14 unknown flow rates and 8 unknown pressure values (at grey nodes). If we now focus on any one node, the flow into and out of the node (pipe junction) must balance — i.e., we have a simple additive equation which contains the flow rates. We can now use the above "friction factor" relationship to replace the flow rates by the respective differences in pressure values driving the flow rates into and out of the node. This results in an equation that links the pressure at the node in question to those at surrounding nodes driving the flow rates. Similarly, all other nodes — 8 in total — will have their own pressure-linking equations. The end result of this process is, therefore, a coupled set of 8 simultaneous equations for the 8 unknown pressure values at the pipe end points. Once this set is solved, we are then able to calculate the flow rates from the "friction factor" relation. Obviously, this is a very simple example, and the analysis of any practical network is performed with elaborate models and computer codes operated by the authorities responsible for the networks.

We rarely, if ever, think about how freshwater reaches our homes, as the entire system of water extraction, processing and distribution is hidden from us. Hopefully, the above description, however brief and superficial, has contributed to an appreciation of how freshwater — "the elixir of life" we take for granted — progresses from evaporation from the ocean surface to our domestic taps.

6.11 Make My Blood Run

We conclude this chapter with a subject close to our heart: blood flow — especially some technological fluid-flow solutions to heart disease. Our body is a machine that runs on blood, air and water. Admittedly, this is a rather facile characterisation of our body's amazingly complex physiology, biology and chemistry. Still, it makes a valid point, reiterating earlier comments in Section 2.3, that our life depends on the flow of many fluid streams that flow through our body, transporting oxygen, CO_2, nutrients, minerals, infection-inhibiting substances, drugs and waste products.

We are all very fortunate to be the beneficiaries of astonishing medical advances over the past decades, which have transformed the quality and length of our lives and which have dramatically eased the pain, discomfort and fear of illness and surgical intervention. The majority of these advances emerged from our ability to unravel bio-chemical mechanisms — the creation of synthetic antibiotic and anti-viral drugs, antiseptics, anaesthetics, analgesics and vaccines being products of this progress. Surgical capabilities have, similarly, made giant steps forward in terms of the range and complexity of the conditions that can now be addressed, aided by advances in imaging technology and medical robotics. The use of radio- and chemotherapy, in combination with surgery, for treating cancer is yet another focal point of medical advances. But what about progress in addressing failures in the fluid-flow functions in our body?

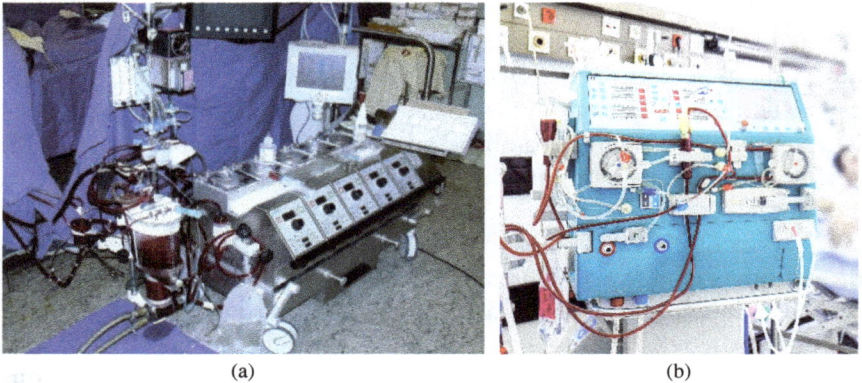

(a)   (b)

Figure 6.47.   Two examples of fluid-flow machines that play crucial roles in medical intervention: (a) a heart-lung machine used in open-heart surgery; (b) a dialysis machine replacing the function of our kidneys.

Source: Shutterstock (with permission); Encyclopaedia Brittanica CC-BY-SA.

Images of modern operating theatres convey strikingly the enormous range and complexity of the equipment supporting surgical interventions, among them fluid-flow machines. Open-heart surgery, in particular, relies heavily on heart-lung (cardiopulmonary bypass) machines, one shown in Figure 6.47. These take over the circulation and oxygenation of the blood, allowing the surgeon to stop the heart, repair it or even replace it with a new heart. Other fluid-flow machines assist respiration and ease the insertion of stents, provide ventricular pumping assistance to the heart, dispense drugs, provide suction of blood and liquids during surgery and emergency treatments, help analyse biological fluids and take over the function of kidneys through dialysis, one machine included in Figure 6.47(b).

Yet, despite this impressive array of advances, one area which has, so far, defied the efforts of technologists is the creation of fully autonomous, long-term operating, implantable devices that replace diseased hearts, kidneys, bladders and lungs. To be clear, there is no dearth of vigorous R&D as well as commercial pressure targeting all these organs, but such machines present especially tough challenges because of a number of engineering and medical obstacles arising from the need for extreme mechanical reliability, the difficulty of managing energy demands and the adverse influence of mechanical actuation (especially shear) on the blood cells. While there have been several cases in which mechanical hearts — two examples shown in Figure 6.48 — have been implanted into patients,

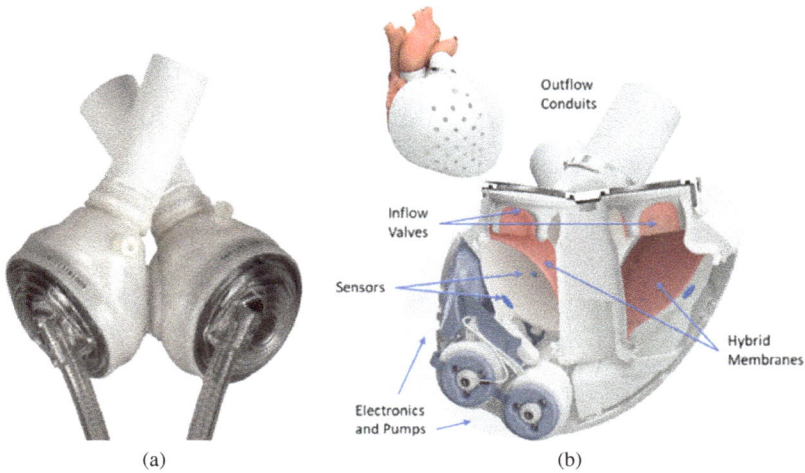

(a)   (b)

Figure 6.48.   Two implantable artificial-heart designs that are used to bridge the removal of the natural heart with the transplant of a new heart: (a) the SynCardia heart containing two pneumatically actuated ventricles; (b) the Carmat heart in which two rotary pumps actuate the oscillatory motion of two membranes.

Source: Slepian *et al.* [41] (with permission); Carpetier *et al.* [42], *The Lancet* and Carmat (with permission).

this was done only as an interim solution to bridge the period between the removal of a diseased heart and the transplant of a healthy human heart. Moreover, the implanted hearts are not autonomous but require external equipment support. The heart shown in Figure 6.48(a), commercially marketed by SynCardia, is a pneumatically driven, pulsatile system that contains two polyurethane ventricles with tilting-disc valves. The design in Figure 6.48(b), marketed by Carmat, is driven by two electrical rotary pumps that actuate two membranes in an oscillatory manner simulating the systole and diastole.

Another difficult R&D battle rages around the realisation of an implantable artificial kidney. Figure 6.49 includes an image of an experimental renal cell implanted into a pig. In contrast to the mechanical heart, the principal challenge here is not mechanical but more of a bio-chemical nature, the only important fluid-mechanic issue being the distortion of the flow structure in the U-shaped blood channel and its effects on the shear strain.

Against the above rather sober account of the current status of *in vivo* implantable fluid-flow devices, a distinctive success story is the, now,

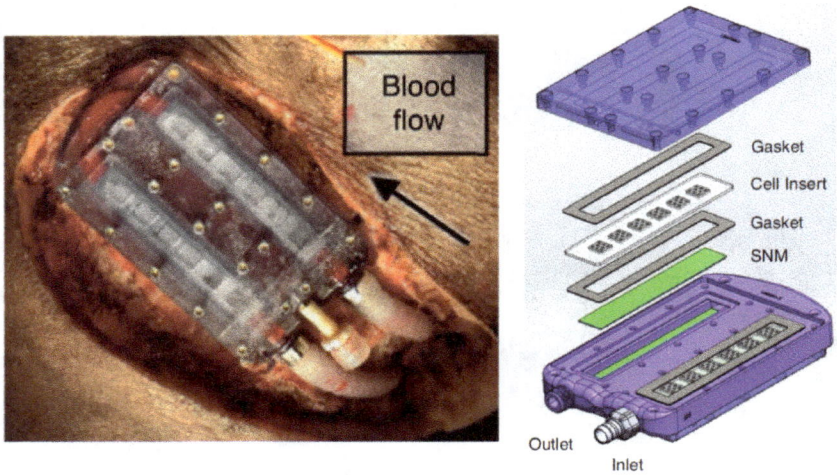

Figure 6.49.   Artificial kidney implanted into a pig.

Source: Kim *et al.* [43] (with permission).

routine artificial replacement of diseased heart valves by artificial ones, exemplified by Figures 6.50 and 6.51. Some introductory comments on such valves were already made in Section 2.3. Here, we expand on these comments by describing key fluid-mechanic features in more detail.

Figure 6.51 shows the four major mechanical heart-valve designs that are used in practice. By way of context, Figure 6.50 shows, schematically, the flow of blood through various parts of the heart, especially its valves, and how a prosthetic heart valve is positioned between the aorta and the lower heart ventricle.

Clearly, the most important requirement of a heart valve design is longevity, for there can be no more catastrophic failure than the disintegration of the valve mechanism or even the detachment of a key component. Longevity is enhanced by simplicity, but there is also the need to minimise the adverse fluid-mechanic effects on the blood cells, as is explained below. A third requirement is the ease of surgical implantation, but this issue is a common concern with all valve designs.

In Figure 6.51, the two right-hand-side columns indicate, respectively, the qualitative flow patterns that arise when the valve is open, in forward flow, and closed, when reverse flow is supposed to be blocked. None of the valves blocks the flow fully, and there is always some undesirable reverse flow leakage. For the bileaflet and tilting-disc designs, two views

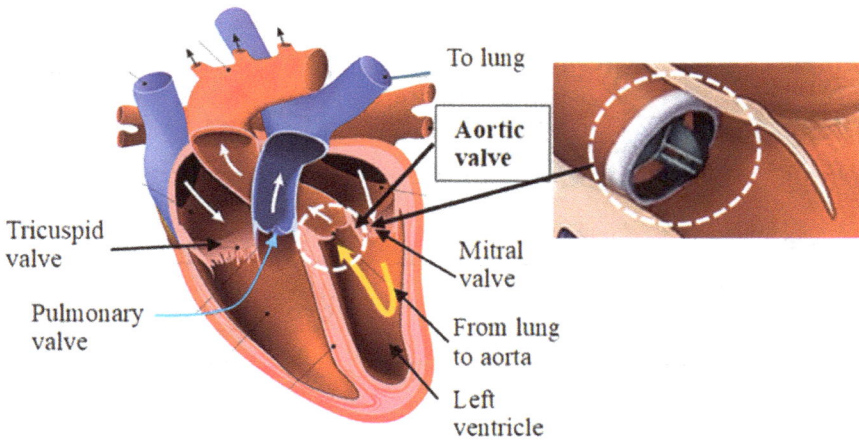

Figure 6.50. Flow of blood through the heart and its four valves, and the position of an implanted artificial heart valve (circled white) controlling the flow from the lungs via the left ventricle to the aorta.

Source: iStock, modified (with permission).

are given: one from the side, along the tilting axes, and the other from the top. All valves contain a rim that is used by the surgeon to connect the valve to the tissue of the heart, and the result is a shoulder, or step, that causes a recirculation region to form in the cavity behind the shoulder. The problem with any recirculation zone is that it encourages the deposition of substances in the cavity and favours the formation of clots and ensuing thrombosis. More serious, however, is the fact that the hard mechanical surfaces of the moving discs generate intense shear layers, both on the hard surfaces and in the regions between the fast central blood jets and the slow flow close to the aorta wall. This shear causes damage to the red blood cells and the platelets, and it also promotes turbulence that has additional undesirable effects on the smooth flow of the blood stream. The tilting-disc design, in particular, favours the formation of large unsteady recirculation zones behind the disc and an asymmetric jetting downstream of the valve.

Of the designs shown in Figure 6.51, the bileaflet valve is the one most widely used in practice. While not being ideal, it facilitates a reasonably well-conditioned forward flow. Although this design also causes the formation of two recirculation regions when open, these are fairly weak.

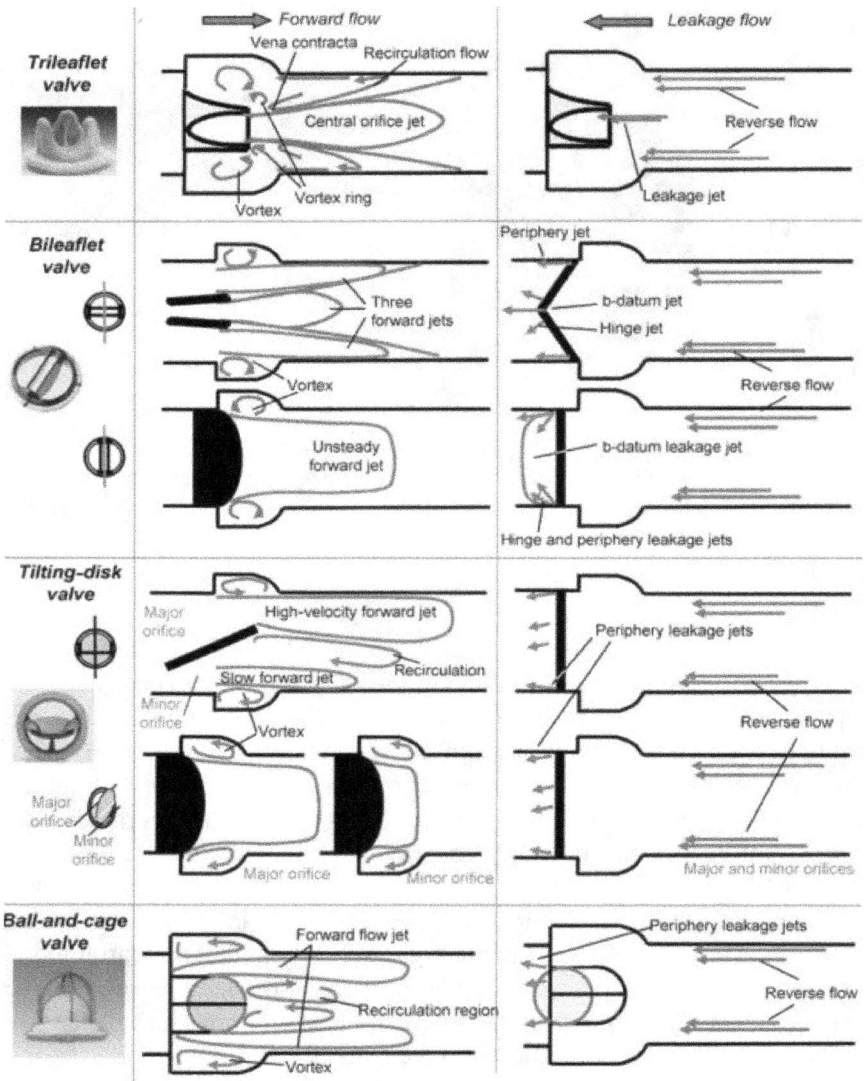

Figure 6.51. Four alternative heart-valve designs showing qualitative flow patterns in open position (middle column) and closed position (right column).

Source: Dasi *et al.* [44] (with permission).

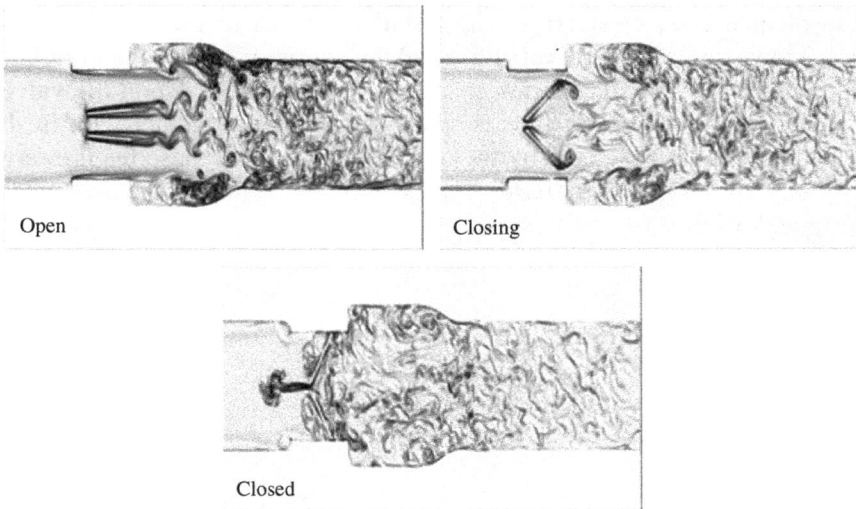

Open

Closing

Closed

Figure 6.52. Snapshots from a computational simulation of an opening and closing artificial bileaflet valve.

Source: Min Yun *et al.* [45] (with permission).

Measurements in heart-valve flows are very difficult, and it is much easier to use flow-simulation software to gain a detailed view of how the flow behaves during the rapid response of the leaflets in systole and diastole. The nature of this software is outlined in Chapter 3 (Figures 3.3–3.5), and variants of such software are equally applicable, in principle, to cars, aircraft, atmospheric flows and heart valves, or indeed any other flow.

Figure 6.52 gives three snapshots derived from a movie that covers the behaviour of a bileaflet valve through the entire heartbeat cycle. The three images relate, respectively, to the fully open position, the closing phase and the fully closed position. As the valve opens, two shear layers are formed at the edges of the leaflets. These become wavy and unstable (see Figure 5.3 in Section 5.3) prior to transitioning to the turbulent state. In the open position, the recirculation zones in the rim cavity of the valve can be identified. As the valve closes, the forward flow slows down and shear layers reduce in size and intensity. In the fully closed state, there is some reverse leakage between the two leaflets as well as between the leaflets and the aorta wall. While the flow behind the closed leaflets is almost quiescent, on average, the turbulent eddies in the fluid weaken only slowly, by viscous dissipation, and the continuing presence of turbulence is clearly seen in the lower image in Figure 6.52. Upon the valve opening again, this turbulence is swept downstream together with fresh turbulence being generated by the opening leaflets. All these features are virtually absent in natural tissue valves, and this illustrates the unavoidable penalties of using artificial valves.

Chapter 7

Command and Control

Having gone, in previous chapters, through the description and interpretation of a wide range of flow configurations in the context of technological scenarios, we return to a central question posed in the introductory Chapter 1, to which only a glib answer was given at that stage:

Why would we want to delve into the inner mechanics of the many flows in which we are emersed and the somewhat fewer ones which are within us? The answer is that, without understanding the mechanisms and processes that are driving these flows, we cannot hope to optimally harness, manipulate and control them...

Although the meaning and relevance of the terms "control" and "manipulation" have not been spelt out, we have, in fact, encountered both in a number of flow configurations discussed in the preceding chapters, as will become clear in the following.

"Flow control" is a collective term for a range of passive and active measures designed to force the flow in question to adhere to particular constraints and/or adopt particular characteristics or properties. Passive methods require no actuation or energy input, while active methods do. The challenge in the latter group is to achieve the desired state at an acceptably low energy input and installation penalties. It turns out that this challenge is very tough for a large proportion of flows to which active control is applied, subject to practical engineering constraints. This partly reflects the fact that flows are extremely "reluctant" to respond favourably to attempts that force them to move away from their natural default state towards the desired controlled state.

We have already encountered a number of control methods without resorting to the term "control". One example is the use of a spiral on the outside of a chimney, as shown in Figure 4.31. Another is the deliberate shape changes introduced in tall buildings to avoid vortex shedding and flow-induced vibrations, also exemplified by Figure 4.31. A third is the deliberate introduction of swirl to stabilise combustion and enhance mixing within the cylinder of an internal combustion engine, as shown in Figures 6.18 and 6.19. A fourth example is the addition of a bulbous nose to large ships to reduce wave drag, as mentioned in Section 6.7 and also seen in Figure 7.8(a) to follow. Finally, we have considered the effects of car spoilers in Section 4.6 (see Figure 4.29), designed to counteract the lift force at high speeds.

In the following three sections, we consider a range of specific flow-control methods that target the reduction of drag and thus energy losses. This is done either by forcing the flow to adhere to the flow-bounding surfaces or by reducing skin friction through controlling the characteristics of the flow close to the bounding surface. The fourth and final section describes strategies, both real and speculative, designed to control the global atmosphere and thus counteract climate change.

7.1 Sleek and Agile

One of the most important means of exercising control in many external flows is to carefully shape the confining or bounding surfaces in a manner that minimises the consequences of undesirable flow behaviour. The most obvious, and almost always most effective, type of flow control is streamlining a body that is moving in a fluid at a high speed and for which we want to reduce the drag — classic examples being cars, aircraft, high-speed trains, ships and submarine bodies. A spectacular example of streamlining in surface-transport vehicles is the nose of Japanese Shinkansen trains. There are many (around 20) shapes used on different train generations and even more on experimental trains, two of which are shown in Figure 7.1. In contrast to the nose of an aircraft, the shapes shown are not merely designed to reduce the drag — mostly around 10–15% relative to a blunt nose at high speeds — but also to reduce noise and especially shock-like pressure pulses when the train enters at high

Figure 7.1. Aerodynamically optimised noses of Japanese Shinkansen (bullet) trains: two examples of "flow control" by streamlining.

Source: M. Akihik, Wikimedia Commons CC-BY.

speeds the hundreds of tunnels on Japan's Shinkansen network. Additional constraints that feature in optimised designs include side-wind stability and low lift.

Less spectacular, but nevertheless no less effective, is the use of deflectors on long-distance freight trucks. Such trucks contribute 25–30% of all transport-related emissions, which is roughly three times higher than aviation-produced emissions and around 10% of total global emissions. The scope for streamlining trucks is obviously much more limited than that of high-speed vehicles, such as the trains depicted in Figure 7.1, being highly constrained by cost and operational flexibility factors. However, some relatively simple modifications, such as those shown in Figure 7.2, have been found to reduce drag by up to 25% under realistic road conditions at elevated motorway speeds. There are three separate modifications included in the figure: a roof deflector combined with a filler that bridges the cab and the trailer, a side skirt below the trailer and a boat tail at the back. As shown by the measured flow fields, included in Figure 7.2, the roof deflector avoids flow impingement on the trailer, reduces the intensity of recirculation between the cab and the trailer by about 80% and substantially diminishes the thickness of the boundary layer above the trailer. The boat tail is predicted to reduce the size of the recirculation zone behind the trailer, and this lowers, on its own, the drag by 5–10%.

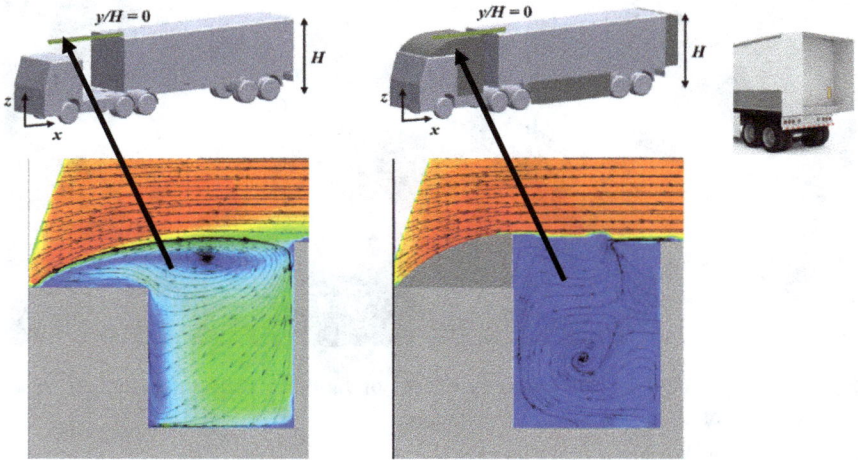

Figure 7.2.   Streamlining of a freight truck by use of a roof deflector, side skirts and a boat tail.

Source: Kim *et al.* [46] (with permission).

As noted already, in relation to both examples above, the optimal shape of a body is often the result of a complex optimisation process in which minimising the drag is, typically, only one of several desiderata. An example is the car body shown in Figure 4.3. It is likely that we could substantially reduce the drag on the body by making its shape similar to that of an elongated drop. However, there are a number of constraints not connected to the drag: the shape should be aesthetically attractive; the cabin needs to be spacious enough to accommodate passengers and their luggage; the lift force has to be low to ensure road stability, including in side winds; the tyres and engine must be accommodated; the rear shape has to be designed such that dirt lifted up by the tyres from the road is not deposited on the rear window; noise and vibrations have to be low; and there are further constraints not included here.

Another example is a wing. Its main function is to generate lift, but at the same time, we want to minimise the drag because this favours low fuel consumption. The two do not go hand in hand: the high-lift-wing configuration shown in Figure 4.28 is an excellent example of flow control targeting maximum lift, but it is also characterised by high levels of drag and intense turbulence-induced noise. A counterexample is the drop-like or fish-like shape of a submarine, which promotes low drag and low acoustic emissions. In particular, the gently narrowing aft body results in a slow adjustment of the boundary layer along the body and inhibits the tendency towards boundary-layer separation (see Figure 4.24).

Is flow control by means of streamlining and shape optimisation for drag reduction purely a dry and dreary engineering-oriented technique? Definitely not — as is strikingly conveyed by Figure 7.3. It shows three closely related examples, taken from the area of competitive non-motorised sport. In all three, winning or losing relies heavily on minimising the drag by careful adaptation of the competitors' posture as they move at high speeds through the air. However, in all three, a careful, drag-conscious design of the equipment is also influential. This is especially so in bike racing, in which the optimum configuration derives from a combination of the posture the rider adopts, the streamlining of the bike components and the helmet and the "engineering" of the fabric and texture of the rider's clothing.

To be at the leading edge of bike racing requires a truly scientific approach that combines wind-tunnel testing with computational modelling of the type used for predicting the flow around aircraft bodies and F1 racing cars (see Figures 3.4 and 3.5 in Section 3.2). The latter is illustrated in Figure 7.3(a) by way of an image derived from a computational study for one particular configuration, in which the colours indicate increments in air speed relative to the speed of the bike and rider — green and blue identifying low-speed and red high-speed regions. An intriguing outcome of this particular computational study, performed for a range of bike designs, is that a replacement of the spokes in both wheels with solid discs yields the minimum drag level. This is by no means obvious because a disc has a much larger surface area than the spokes. The implication of this result is that the increase in friction drag caused by the discs is more than compensated for by the elimination of the turbulent wakes of the individual spokes and the effect of the wakes on the flow around the frame. Figure 7.3 also conveys a remarkably close similarity in the posture adopted by the bike rider and the downhill skier, a shape as close to an elongated, slim drop-like body as the competitor can achieve. In contrast, the ski jumper adopts a wing-like posture, that aims to increase the lift over the jumper's curved back and reduce the drag by the forward tilt of the body, which minimises the area projected against the direction of flight.

(a)

(b)

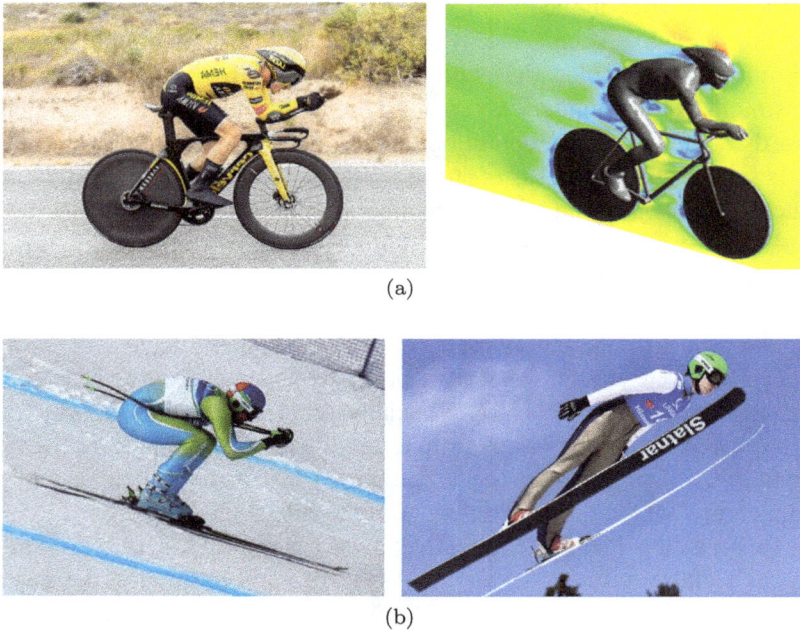

Figure 7.3.   Streamlining in competitive sports: (a) bike racing; (b) downhill skiing and ski jumping.

Source: (a, right) Mannion *et al.* [47] (with permission); (a, left) and (b) Wikipedia CC-BY.

7.2 Clinging to Twists and Turns

In internal flows, we try to avoid steps and excessive area expansions because this causes separation, as shown in Figures 4.20 and 4.32. Separation does not merely distort the flow; it is also a powerful promoter of turbulence in the separated shear layer and the recirculation bubble, with a consequent loss of useful energy. In duct systems, we try to avoid bends and constrictions — e.g., by placing valves in the duct or pipe conveying the flow. Bends generate strong three-dimensional distortions in the flow and possibly regions of separation in high-curvature geometries, thus increasing losses and reducing the pressure energy in the flow. Similarly, a valve or an orifice causes separation and turbulence because it acts like a bluff body within the duct or pipe.

As stated repeatedly, turbulence is undesirable in most circumstances in which energy loss is to be minimised, and many control strategies target its suppression or reduction. What are the exceptions? Three cases have already been pointed to: reacting flows, in which turbulence enhances mixing between the fuel and oxidant (Figure 2.11); dispersion of plumes discharged from chimneys (Figure 2.10); and the avoidance or inhibition of separation on aerofoils and wings. In at least one example — the golf ball in Figure 5.7 — we have seen that indentations can be added to a curved surface, with the objective of delaying separation from the surface.

The last example listed above is a special case of a broader group of methods, which are referred to collectively as "vortex generators". We have already encountered an example that exploits vortex generators: look at Figure 4.36 and focus on the regular, block-like obstacles placed on the bottom wind-tunnel wall upstream of the cityscape model. Here, these obstacles are designed to thicken the boundary layer before the flow reaches the architectural model, so as to more faithfully emulate the thick atmospheric boundary layer found in the natural environment. In the large majority of circumstances, however, vortex generators are used to generate or enhance turbulence in the boundary layer upstream of a region in which the flow is subjected to an adverse pressure gradient and is, therefore, in danger of separating from the wall. Vortex generators come in many shapes and sizes; a few are shown in Figure 7.4. All are, essentially, small obstacles placed upon the surface on which the boundary layer develops. In fact, any protrusion, whatever its shape, acts like a vortex generator. The obstacles cause local, small-scale regions of separation and vortices to form, which enhance the level of turbulence. The consequence is the intensification of mixing near the wall and an increase in the transport of momentum from the outer flow to the near-wall region, which then inhibits separation.

A disadvantage of passive vortex generators is that they cannot be "switched off" when not needed — for example, on a wing when it changes from a high angle of attack during takeoff or landing to cruise conditions in which a low level of drag is crucially important to low fuel consumption and high efficiency. An active control method that can be switched on and off is to inject microjets, either in a steady or periodic fashion, into the flow through holes in the surface, as shown in Figure 7.5. In effect, the microjets are a fluid-flow-equivalent of the static vortex generators. However, this requires a complex, costly and risky actuation system, possibly involving hundreds of jets, which may negate the advantages derived from the flexible, flow-dependent operation. Microjets are

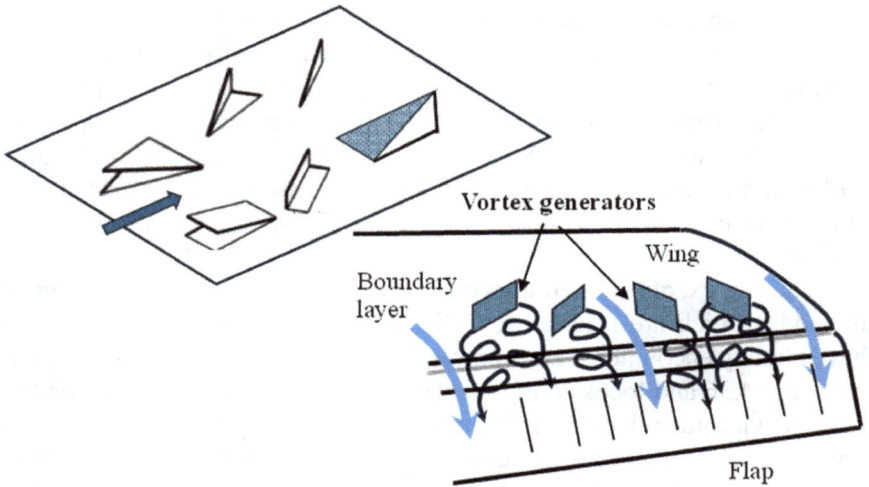

Figure 7.4.   Different vortex generators used to increase turbulence in boundary layers, and one application in a high-lift wing (relative size of vortex generators exaggerated).

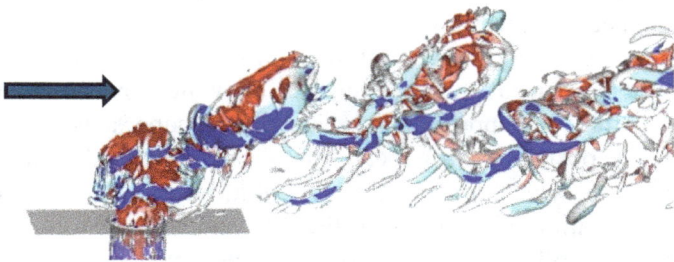

Figure 7.5.   Computational simulation of a microjet injected periodically into a boundary layer for the purpose of increasing turbulence.

also noisy because their discharge velocity needs to be a significant fraction of the speed of the primary flow to be controlled. Periodically actuated microjets are especially noisy because the periodicity causes high-intensity pressure pulses.

   The use of microjets is one of several active control methods that energise the boundary layer and thus inhibit its separation. Two techniques that do not rely on an increased level of turbulence to do so are shown in Figure 7.6. One involves the injection of high-momentum fluid, either steadily or periodically, from a thin, flow-oriented slot at the wall.

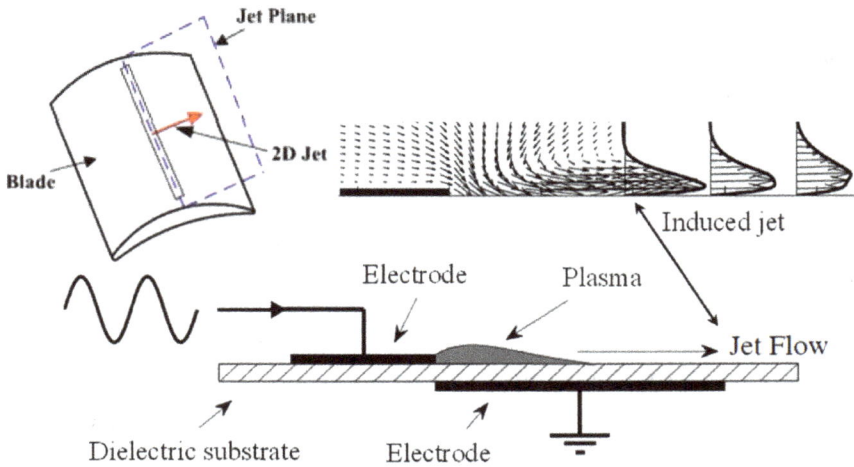

Figure 7.6. The use of flow-directed jet injection and a dielectric-barrier plasma actuator to increase the near-wall momentum in a boundary layer.

Source: K.-S. Choi (private communication).

The other uses a high voltage applied between two electrodes to produce high-momentum, hot plasma pulses, most frequently parallel to the flow. Disadvantages of the latter method are, first, that it requires a risky high-voltage supply (typically 10000–20000 V) and, second, that the actuation requires a significant amount of energy to operate the device.

7.3 Grooving and Moving

A number of passive and active flow-control methods are also employed to suppress the transition to turbulence in a boundary layer and to reduce the friction drag at the wall. Figures 5.4 and 5.8 show that a boundary layer begins its life as a thin shear layer in a laminar (non-turbulent) state and then transitions to a turbulent state through a number of instability modes. The friction drag in a laminar boundary layer is substantially lower than that in a turbulent boundary layer discussed with reference to Figure 5.8. One method that is used to delay transition is to apply suction through a porous portion of the wall on which the control is desired. Suction removes the layer at the wall that is slowed down by friction, thus preventing the boundary layer from thickening and transitioning to a turbulent state. Here again, however, a drawback is that the method requires energy (or work) that may exceed the gain achieved by the delay in transition.

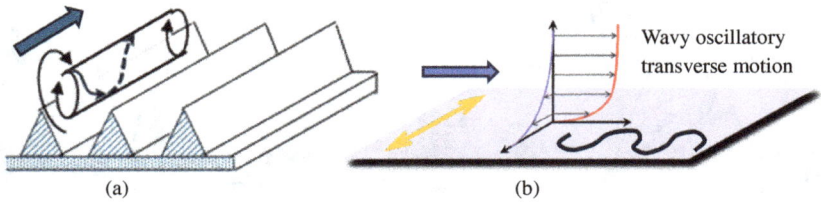

Figure 7.7. Two techniques for reducing the drag in a turbulent boundary layer: (a) riblets and (b) oscillating the wall in the direction normal to the flow (the thick blue arrows indicate the flow direction).

Once a boundary layer has become turbulent, there are only a few methods that allow the drag to be reduced. One passive method is to cover the solid wall with a film that contains longitudinal grooves and riblets, as shown in Figure 7.7(a). The sharp-crested riblets are, typically, only a few microns in height — almost invisible to the naked eye — and have been found to reduce the friction drag by a relatively modest 8–10%.

Although the use of riblets originated in aeronautics, it has also been adopted in the sports arena, specifically for coating the outer layer of swimwear for competitive swimmers. The mechanism at play is not fully understood in all its details. One paradigm is based on the proposition that the raised riblet sides weaken the near-wall turbulence by inhibiting spanwise (normal-to-riblet) fluctuations below the crests, while a second suggests that the drag reduction is rooted, as indicated in Figure 7.7(a), in the formation of vortices within the troughs that "pump" low-momentum fluid from the troughs into the layer just above the crests, thus reducing turbulence in that layer.

Far higher levels of drag reduction, up to 50% in ideal circumstances, can be achieved by a sideways oscillatory actuation of the wall in a direction normal to the boundary layer. One option is to oscillate the whole wall uniformly, but the highest drag reduction is achieved by a wavy motion of different parts of the wall via the streamwise phase shift indicated in Figure 7.7(b). Expressed in highly simplified terms, the oscillatory motion "disrupts" the mechanism by which shear generates turbulence at the wall, thus depressing the turbulence intensity. This method is only conceptual in nature, however, and is extremely difficult to realise in practice. In any event, the energy needed to actuate the wall can easily exceed that saved by the reduced drag in the boundary layer, although computational simulations indicate that there are parametric regimes within which a net gain in energy is possible.

When liquids are transported over long distances, the total friction force in the pipeline can be very high, and this may require the use of extremely high-powered pumps. An effective drag-reduction technique, discovered in the 1960s and extensively explored in the 1970s [50], involves the injection of small amounts of long-chain-molecule polymers into the transported liquid. Intriguingly, even at very low concentrations, these polymers are observed to reduce drag by as much as 80%. They do so by "absorbing" and damping the turbulence energy in the sheared pipe flow and inhibiting the generation of fresh turbulence close to the wall. However, the drag-reduction effectiveness of the polymeric liquids tends to decline steadily over time when sheared, requiring a continuous supply of the polymer — a costly and difficult process in practice.

A final drag-reduction technique discussed here is the injection of air bubbles into a liquid boundary layer. This concept has been investigated, both theoretically and experimentally, for over a decade in ship hulls, as illustrated in Figure 7.8(a). The rationale is that the air bubbles close to the

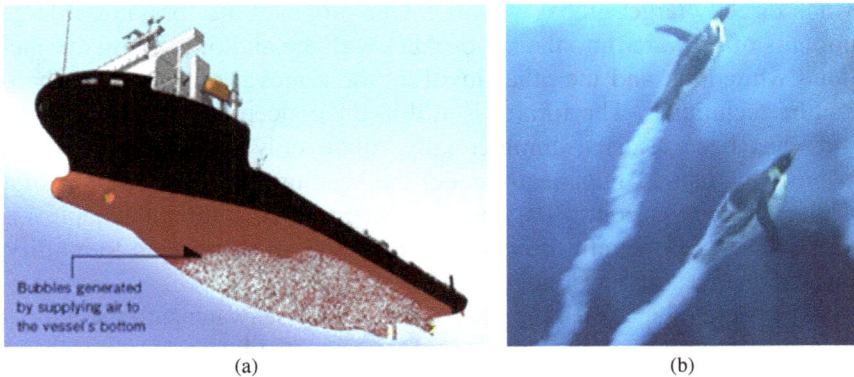

(a)   (b)

Figure 7.8.   Friction drag reduction by means of an air-bubble film on the outer skin of bodies propelled in a marine environment: (a) Mitsubishi's concept of injecting air bubbles at the bottom of a ship; (b) Emperor Penguins releasing drag-reducing air bubbles from their plumage during a high-speed dive towards an ice float.

Source: (a) Kawakita *et al.* [48], Mitsubishi Heavy Industries (with permission); (b) The Blue Planet, BBC (with permission).

wall create a low-viscosity slip layer — an air cushion — that reduces friction drag. However, the practical realisation of the concept is challenging. Specifically, the technique is difficult to implement from a structural point of view, requiring the wall to be perforated or equipped with external injection nozzles, is energetically expensive and tends to be compromised by the buildup of barnacles on the ship hull, which can increase the drag by a much higher margin than the reduction achieved by the air-bubble control.

An interesting aspect of the last method discussed above is that it represents one of several attempts to emulate flow-control solutions observed in nature, with almost all aiming to reduce skin-friction drag. As is shown in Figure 7.8(b), penguins diving into water carry with them air in their dense feathers into the water, releasing the air as a bubbly film, thus reducing the drag on their bodies and allowing them to manoeuvre with extreme agility and speed through water, which is especially advantageous when they jump out of the water onto slippery ice floats.

7.4 Saving the Planet?

We close this chapter with remarks on two *"grand-challenge"* geo-engineering projects in which fluid flow plays the principal role and which directly target global warming. One proposes the spraying of large amounts of seawater into the atmosphere with the aim of creating marine-cloud whitening, and the other involves the removal of CO_2 by carbon-capture techniques. The former is still in the conceptual stage, while the latter has been realised commercially, albeit only to a modest extent. Admittedly, neither falls neatly under the heading *"flow control"*, insofar as neither involves controlling the flows themselves. Rather, the control is applied to the global atmosphere by modifying, or attempting to modify, adverse changes in our climate.

The logic of spraying large amounts of salty seawater into the atmosphere is rooted in the observation that soot emitted by ships crossing the oceans encourages the formation of white marine clouds, which reflect sunlight back into space. To realise the formation of white clouds over large areas requires the release of very large amounts of salty water into the atmosphere above the oceans, with salt particles acting as seeds for water condensation at high altitudes. The concept shown in Figure 7.9 involves unmanned wind-driven ships crisscrossing the oceans and continuously spraying large amounts of water into the atmosphere. Little has been done beyond small-scale testing towards realising this ambitious

Figure 7.9. Concept of spraying seawater into the atmosphere to promote the formation of white clouds that reflect sunshine away from Earth.

Source: © J. MacNeill (with permission).

concept. It would involve addressing numerous fluid mechanics problems, including the wind-propulsion system based on circular "sails" or long spinning cylinders, called "Flettner rotors" (note the connection to wind-induced vibrations discussed in Section 4.7), the power-generating turbines below the hull, the transport of seawater into the spray tubes, the design and application of the spray-generating nozzles and the behaviour of jet-like spray as it issues into the atmosphere and is being transported to cloud-forming layers. There is broad consensus that, even if all technological challenges can be overcome and the effects on the weather are understood, this technique can, at best, only be a temporary solution and will only be of limited effectiveness if set against the costs involved and the magnitude of the processes that contribute to global warming.

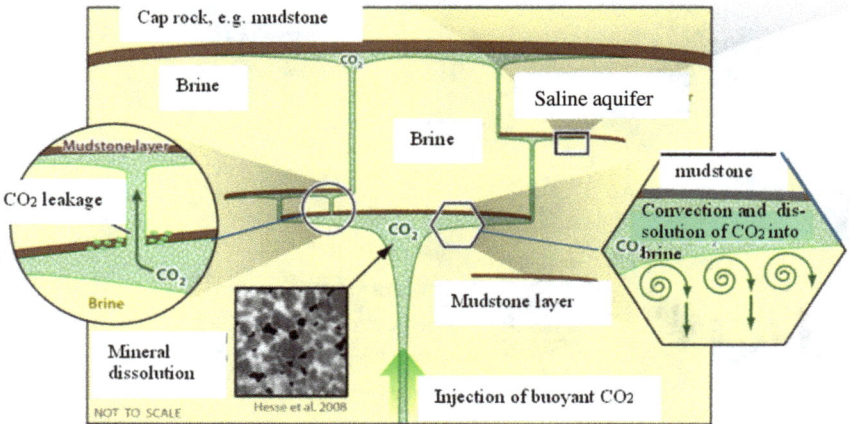

Figure 7.10. Schematic representation of some of the fluid mechanics challenges associated with underground storage of sequestrated CO_2.

Source: Huppert and Neufeld [49] (with permission).

The fluid-mechanics challenges involved in CO_2 sequestration, combined with underground storage, are complex and multifaceted. One of them concerns the extraction and transportation processes. The sequestration itself takes place, most frequently, as an adjunct to power generation or chemical processing in industrial plants. However, a far more difficult aspect is the interaction of the CO_2 with the geology of the underground reservoirs into which it is being pumped. A reservoir is not simply an empty space, but a complex system of layers of rock, mud, soil, shale, coal, brine and saline aquifers.

Figure 7.10 provides a schematic summary of some of the fluid-flow issues that need to be addressed. CO_2 is absorbed by the brine and is transported with it by diffusion and/or buoyancy-induced convection; it leaks through channels in the rock and rises towards a mudstone or rock cap covering the reservoir. Hence, any attempt to model the interactions — modelling being the only realistic approach to gaining insight into what happens in the reservoir — is a difficult and uncertain exercise. It relies on geography-specific seismic imaging, the use of approximate mathematical models to describe the dissolution of CO_2 in brine, the containment of CO_2 within various geological formations and the trapping, leakage and upward transport of CO_2.

Another type of carbon-capture strategy is to extract CO_2 directly from the atmosphere — i.e., unrelated to any particular industrial

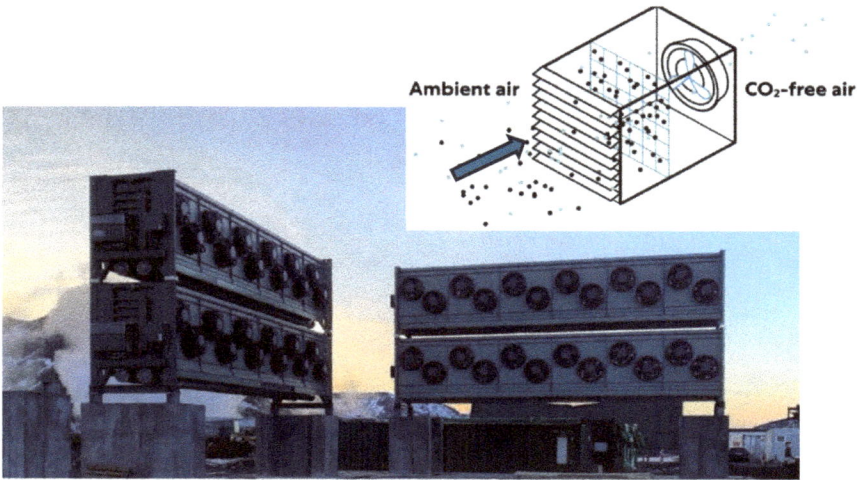

Figure 7.11. A commercial plant for direct air extraction of CO_2 from the atmosphere.
Source: © Climeworks AG (with permission).

process — and store it in the form of fertilisers and other synthetic substances or pump it, dissolved in water, into underground storage cavities, preferably of a type in which the CO_2 reacts with minerals to form a rocklike substance over long periods of time.

The principles of the extraction process are shown in Figure 7.11. Air is sucked by dozens of large fans into plenum chambers equipped with CO_2 filters, onto which the CO_2 is chemically deposited. Once a filter is saturated, heat is used to liberate the CO_2 from the filter and either embed it into solids or transport it to an underground storage area. As the process requires substantial amounts of energy to drive the fans and to separate the CO_2 from the filters, direct air extraction is only economically tenable where low-cost electrical and thermal energy is available — the best option being geothermal energy readily available — for example, in Iceland, NZ, Japan and other volcanically active geographical locations.

The largest plant in operation, as of 2024, was constructed by the Swiss company Climeworks in Iceland and extracts around 4000 tons of CO_2 per year from the atmosphere. This is a relatively modest amount, but the plant shows a promising route that could significantly contribute to halting or at least slowing global warming.

Epilogue: What Else Remains to Be Said?

Despite the dozens of flow configurations discussed in the preceding chapters, the answer to the question posed in the above title is — unfortunately, from my perspective — *"a great deal"*.

The first paragraph of Chapter 6, preceding 10 Sections in which a broad range of thematically different flow scenarios were discussed, contains the following sentence:

> *Yet, these are but a tiny fraction of the enormous number of environmental systems and man-made machines and processes which govern our lives and on which we depend daily.*

Notwithstanding the many flows introduced in Chapter 6, and many more discussed in other parts of this book, my resigned response remains valid.

There are hundreds of further natural phenomena and engineering applications that arguably merit being included. Under the former heading, little if anything has been said about ocean currents — the Gulf Stream, for example — and (very slow) magma flows under the Earth's crust. Flows at this global scale are extremely complicated by the multifacetted interactions among many parameters and physical mechanisms. In the case of oceanic circulations, the nature of the complexity involved was hinted at in Section 2.5. Similarly, only some atmospheric manifestations of volcanism were outlined in Section 2.4. However, what happens below the Earth's surface was left unsaid, mostly because magma flows are virtually impossible to describe in terms that are reasonably transparent and general, even if one focuses only on the flow between an idealised magma reservoir and the volcanic plume. Magna is a mixture of solid and molten rocks at high pressure within which gases are dissolved. As the magma rises, the gas is expelled from the liquid or solid mixture in the form of growing bubbles, eventually leading to the pre-plume flow that is a fragmented mix of solids, liquid, gas, ash and steam, all flowing through a geologically convoluted geometry that is specific to the geographic location of the volcano being considered. The flow of magma below the Earth's crust is akin, in some respects, to oceanic currents but is far slower, has very different physico-chemical properties and is obstructed by the complex underground topography of the interface between the crust and the liquid magma. Underground water streams and aquifers — yet another class of flow not discussed more than superficially in the context of hydrology, in Section 6.10 — are less complicated, but still do not lend themselves to a transparent general discussion, because they are confined within a huge variety of complex geological structures, an idealised impression of which is conveyed by Figure 7.10.

The complex bio-mechanics by which fish, seals, dolphins, sharks and whales swim at high speed is another area that is worthy of exploration. This may be extended to flying fish and diving birds, supplementing the remarks on jumping penguins included in Chapter 7. In many of these animals, the scaly and feathery structures on the skin play a crucial role in reducing the drag during swimming that allows the animals in question to move so gracefully and at such high speed through the water. The scaly skin of fish and sharks and the particular hairy structure on the skin of seals are assumed to condition the near-wall boundary layer in a manner that reduces the drag relative to that on a flat surface. However, the drag-reduction mechanisms are nowhere near well established or understood, and the presence of slimy or oily bio-fluids on the skin obscures the role of fluid mechanics — in particular, the manner in which the intensity of turbulence close to the skin might be reduced in response to the geometric features on the skin surface.

While basic aerodynamic aspects of birds' flight were described in Section 4.5, albeit in rather superficial terms, the flight of insects was ignored. Insects do not flap their wings in a manner birds do but oscillate their wings up and down while deforming them to generate lift and forward motion. These are some of the astonishing biological solutions in the natural world, and engineers are often trying to emulate such natural solutions, usually reporting their efforts under headings starting with *"bio-inspired..."*. A rather crude emulation is the use of riblets on surfaces (see Figure 7.7, Chapter 7) — e.g., on aeroplane wings or competition swimwear — designed to reduce friction drag. Another *bio-inspired* flow, covered in Chapter 7, Figure 7.8, is the injection of air bubbles through submerged ship hulls. However, all these bio-inspired solutions are rarely effective, if only because the emulation is simplistic, energetically inefficient — if involving active actuation — and ignores the complexities of the control which animals are able to impart on the flows around their bodies. For example, insect-like micro UAVs (unmanned aerial vehicles) with oscillating or flapping wings have been designed and made to fly, but these are not efficient relative to quadcopters (drones) and suffer greatly from power and structural limitations.

In the engineering sphere, topics that might deserve to be included, or described in greater detail than done in this book, are oil and gas extraction, petrochemical processing, flood prevention, water purification, liquid-food processing and packaging, ink-jet and metal printing, drug production, medical diagnostics based on the manipulation of bodily fluids and the aerodynamics of high-speed trains (see Figure 7.1), and so it goes on and on.

On each and every one of the topics and areas discussed or mentioned in this book, numerous scientific papers and, in many cases, thick textbooks have been published — although it is appropriate to add that flow-physical issues may not be pivotal or especially interesting in all cases. For example, helicopter rotors have been mentioned only briefly in Chapter 6 as a particular form of a propeller. Yet, a recent book by J.G. Leishman on the *Principles of Helicopter Aerodynamics* (Cambridge University Press, ISBN 978-0-521-85860) has over 800 pages full of mathematical analysis and graphs. Basic aspects of heat transfer in nuclear reactors were also discussed briefly in Chapter 6. Yet, a book by R.E. Masterson on *Nuclear Reactor Thermal Hydraulics* (CRC Press, 978-1-138035379) has almost 1400 pages.

Dozens of companies are engaged in the design, development, manufacture and exploitation of products and processes to which fluid flows are central, hundreds of numerical analysts and software engineers are beavering away in software companies and universities, developing and applying computational simulation methods for an ever-broadening range of flow-physical processes, and many research organisations and universities operate large wind tunnels, some up to 35×25 metres in cross-section, to validate computational predictions and to examine realistic models of buildings, aircraft, trains and cars.

An early title of this book was: *What Have Fluids Ever Done for Us?* I claim that we have very good reason to answer this question with *"an immeasurable amount"*, for the book conveys the fact that we are immersed in innumerable fluid flows and hundreds of engineering solutions that underpin modern life as we know it.

As to the era-defining issue of global warming and climate change, we know that there is no magic bullet to remove this monumental threat to mankind. In fact, rather depressingly, it is widely accepted that, even if we stop all CO_2 emissions today, the CO_2 content in the atmosphere and global temperatures will stabilise and decline extremely slowly, over hundreds of years. At best, we can only hope to slow down the process. Fluid mechanics, in a technological context, plays a pivotal role in our efforts to do so. Indeed, understanding climate change itself and the global processes driving it relies critically on insight into the phenomena governing atmospheric and oceanic flows. Hopefully, this book has contributed, if only modestly, to a wider appreciation of the importance of fluid flows to our present life and our future existence on this Earth.

Acknowledgements

As already mentioned in the Preface, my granddaughter Maya provided the primary inspiration for this book by challenging me to write a guide to "fluid mechanics for dummies". I cannot claim to have risen to her challenge — insofar as this book is meant for scientifically minded readers rather than "dummies". Nevertheless, Maya deserves to be acknowledged, here again. Otherwise, writing this book was, pretty much, a solitary endeavour, though one crucially encouraged and supported by my wife, Freda, who has provided me with the comforts of a home run with military efficiency. All those I acknowledge in the following helped me *after* the book was largely completed.

To my surprise — and dismay — the process of securing permissions for including images and photos — those not my own — turned out to be a stressful and, in parts, dispiriting experience. In fact, had I known the scale of the challenge — the amount of time expended, the close to 1000 emails sent out, many remaining unanswered even after repeated reminders, the disappointment of being passed from pillar to post and the outrageous charges demanded for even the simplest of images — I can honestly say that I would have never embarked on this project. Let this be a warning to any prospective author who may wish to embark on writing a book that is to include more than a handful of images imported from outside sources.

Pleasingly, at the opposite end of the scale, some colleagues I approached stood out as being especially responsive and helpful in facilitating the process of securing permissions. A few have provided me with unpublished or even freshly prepared images. Others have prodded unresponsive sources on my behalf, calling in favours. I hope all those who granted me permission are formally acknowledged in the captions of relevant figures.

Those I wish to single out here for their generous support are: Prof. Hiroyuki Asada (Tohoku University), Ms. Neena Banlawe (COMSOL), Prof. Michael Breuer (Helmut Schmidt University), Dr. Song Chen (Singapore National Environmental Agency), Prof. Kwing-So Choi (Nottingham University), Mr. Tim Davie (DG, BBC), Prof. Rudolf Dvořák (The Czech Academy of Sciences), Prof. Koji Fukagata (Keio University), Prof. Herbert Huppert (Cambridge University), Dr. Sylvain Lardeau of (Siemens PLC), Prof. Paul Markowski (Pennsylvania State University), Dr. Florian Menter (ANSYS), Prof. Ramis Örlü (KTH), Prof. Sjoerd Rientra (Eindhoven University of Technology), Prof. Maria Vittoria Salvetti (University of Pisa), Prof. Harry Schulz (Auburn University), Prof. Kazuhiko Suga (Osaka Metropolitan University), Dr. Zhong-Nan Wang (Birmingham University), Prof. Ajit Yoganathan (Georgia Institute of Technology), Dipl. Ing. Mathiew Weber (Magma GmbH) and Nick Zentil (Titan Metal Fabricators). My sincere thanks go to all of you, and my apologies go to those who I have, either inadvertently or unjustly, left out.

References

[1] Liu, Y., Kojima, T., and Fujita, Y. (2003). Study on structure of pseudo-shock waves in under-expanded jet. *Journal of Visualization Society of Japan*, 23, 53–56.

[2] Settles, G. S. (2001). *Schlieren and shadowgraph techniques: Visualizing phenomena in transparent media.* Springer-Verlag.

[3] Syed, F., Khan, S., and Toma, M. (2023). Modeling dynamics of the cardio-vascular system using fluid-structure interaction methods. *Biology*, 12, Article 1026.

[4] Mishra, K. B., Wehrstedt, K. D., and Schoenbucher, A. (2010). Safety aspects of organic peroxide pool fires. *Proceedings of Combura 2010*.

[5] Correia, L. P., Rafael, S., Sorte, S., Rodrigues, V., Borrego, C., and Monteiro, A. (2021). High-resolution analysis of wind flow behavior on ship stacks configuration: A Portuguese case study. *Atmosphere*, 12, Article 303.

[6] Ezzddine, W., Schulz, J., and Rezg, N. (2019). Pitot sensor air flow measurement accuracy: Causal modelling and failure risk analysis. *Flow Measurement and Instrumentation*, 65, 7–15.

[7] Asada, H., Tamaki, Y., Takaki, R., and Kawai, S. (2023). FFVHC-ACE: Fully automated Cartesian-grid-based solver for compressible large-eddy simulation. *AIAA Journal*, 61, 3466–3484.

[8] Sivaraj, G., Parammasivam, K. M., Prasath, M. S., Vadivelu, P., and Lakshmanan, D. (2021). Flow analysis of rear end body shape of the vehicle for better aerodynamic performance. *Materials Today: Proceedings*, 47, 2175–2181.

[9] Dabiri, J. O., Howland, M. F., Fu, M. K., and Goldshmid, R. H. (2023). Visual anemometry: Physics-informed inference of wind for renewable

energy, urban sustainability, and environmental science. *arXiv Physics*, Article 2304.04728.

[10]   Mariotti, A., Buresi, G., and Salvetti, M. V. (2014). Control of the turbulent flow in a plane diffuser through optimized contoured cavities. *European Journal of Mechanics — B/Fluids*, 48(4–5).

[11]   Leschziner, M. A. (1995). Computation of aerodynamic flows with turbulence-transport models based on second-moment closure. *Computers and Fluids*, 24, 377–392.

[12]   Ghouila-Houri, C., Gallas, Q., Garnier, E., Merlen, A., Viard, R., Talbi, A., and Pernod, P. (2017). High temperature gradient calorimetric wall shear stress micro-sensor for flow separation detection. *Sensors and Actuators A*, 266, 232–241.

[13]   Dvořák, R. (2016). Aerodynamics of bird flight. *EPJ Web of Conferences*, EFM15, 114, Article 01001.

[14]   Sereez, M., and Zaffar, U. (2021). Dynamic stall on high-lift airfoil 30P30N in ground proximity. *Open Journal in Fluid Dynamics*, 11, 135–152.

[15]   Griffin, O. M., and Ramberg, S. E. (1974). The vortex street wakes of vibrating cylinders. *Journal of Fluid Mechanics*, 66, 553–576.

[16]   Biswas, G., Breuer, M., and Durst, F. (2004). Backward-facing step flows for various expansion ratios at low and moderate Reynolds numbers. *Journal of Fluids Engineering*, 126(3), 362–374.

[17]   Breuer, M., Bernsdorf, J., Zeiser, T., and Durst, F. (2000). Accurate computations of the laminar flow past a square cylinder based on two different methods: Lattice-Boltzmann and finite-volume. *International Journal of Heat and Fluid Flow*, 21, 186–197.

[18]   Yakhot, A., Liu, H., and Nikitin, N. (2006). Turbulent flow around a wall-mounted cube: A direct numerical simulation, *International Journal of Heat and Fluid Flow*, 27, 994–1009.

[19]   Breuer, M., Lakehal, D., and Rodi, W. (1996). Flow around a surface mounted cubical obstacle: Comparison of LES and RANS. *Notes on Numerical Fluid Mechanics*, Springer, 53, 22–30.

[20]   Cao, Y., Tamura, T., Zhou, D., and Bao, Y. (2021). Topological description of near-wall flows around a surface-mounted square cylinder at high Reynolds numbers. *Journal of Fluid Mechanics*, 933, Article A39.

[21]   Zhang, S., Jaworski, A. J., McParlin, S. C., and Turner, J. T. (2019). Experimental investigation of the flow structures over a 40° swept wing. *The Aeronautical Journal*, 123, 39–55.

[22]   Markowski, P., and Richardson, Y. (2014). What we know and don't know about tornado formation. *Physics Today*, 67, 26–31.

[23]   Kok, J. C., and van der Ven, H. (2012). *Capturing free shear layers in hybrid RANS-LES simulations of separated flow* (Report No. NLR-TP-2012-333).

[24] Veerasamy, D. (2019). *Effect on Flap Transition of Upstream Wake Turbulence* [Doctoral dissertation, City University London].

[25] Chen, C., and He, L. (2023). Two-scale solution for tripped turbulent boundary layer. *Journal of Fluid Mechanics*, 955, Article A5.

[26] Luelff, J. (2015). *Statistical and Dynamical Properties of Convecting Systems* [Doctoral dissertation, University of Muenster].

[27] de' Michieli Vitturi, M., and Pardini, F. (2021). PLUME-MoM-TSM 1.0.0: A volcanic column and umbrella cloud spreading model. *Geoscientific Model Development*, 14, 1345–1377.

[28] Intergovernmental Panel on Climate Change. (1999). *IPCC report on aviation and the global atmosphere*. Cambridge University Press.

[29] Wang, Z. N., Tyacke, J., and Tucker, P. (2018). Large eddy simulation of serration effects on an ultra-high-bypass-ratio engine exhaust jet. *Comptus Rendus Mechanique*, 346, 964–977.

[30] Redonnet, S., Ben Khelil, S., Bulté, J., and Cunha, G. (2016). Numerical characterization of landing gear aeroacoustics using advanced simulation and analysis techniques. *Journal of Sound and Vibrations*, 403, 214–233.

[31] Chen, S., Gojon, R., and Mihaescu, M. (2021). Flow and aeroacoustic attributes of highly-heated transitional rectangular supersonic jets. *Aerospace Science and Technology*, 114, Article 106747.

[32] Groot, J. A. W. M. (2011). *Numerical Shape Optimisation in Blow Moulding* [Doctoral dissertation, University of Technology Eindhoven].

[33] Cleary, P., Ha, J., Alguine, V., and Nguyen, T. (2002). Flow modelling in casting processes. *Applied Mathematical Modelling*, 26, 171–190.

[34] Cleary, P., Ha, J., Prakas, M., and Nguyen, T. (2010). Short shots and industrial case studies: Understanding fluid flow and solidification in high pressure die casting. *Applied Mathematical Modelling*, 34, 2018–2033.

[35] Thomas, B. G. (2003). Modeling of continuous casting. The AISE Steel Foundation.

[36] Fenton, J. D. (1988). The numerical solution of steady water wave problems. *Computers and Geosciences*, 14, 357–368.

[37] Earle, S. (2019). *Physical geology*. B.C. Open Collection by BCcampus.

[38] Link, O., Mignot, E., Roux, S., Camenen, B., Escauriaza, C., Chauchat, J., Brevis, W., and Manfreda, S. (2019). Scour at bridge foundations in supercritical flows: An analysis of knowledge gaps. *Water*, 11, Article 1656.

[39] Schulz, H. E., Dorn Nubrega, J., Andrade Simoes, A. L., Schulz, H., and de Melo Porto, R. (2015). Details of hydraulic jumps for design criteria of hydraulic structures. In *Hydrodynamics concepts and experiments* (Chapter 4). IntechOpen.

[40] Zhang, Q., Zheng, F., Jiu, Y., Savic, D., and Kapelan, Z. (2021). Real-time foul sewer hydraulic modelling driven by water consumption data from water distribution systems. *Water Research*, 188, Article 116544.

[41] Slepian, M. J., Alemu, Y., Soares, J. S., Smith, R. G., Einav, S., and Bluestein, D. (2013). The Syncardia total artificial heart: In vivo, in vitro, and computational modeling studies. *Journal of Biomechanics*, 46, 266–275.

[42] Carpentier, A., Latrémouille, C., Cholley, B., Smadja, D. M., Roussel, J-C., Boissier, E., Trochu, J. N., Gueff, J. P., Treillot, M., Bizouarn, P., Méléard, D., Boughenou, M. F., Ponzio, O., Grimmé, M., Capel, A., Jansen, P., Hagège, A., Desnos, M., Fabiani, J-N., and Duvea, D. (2015). First clinical use of a bioprosthetic total artificial heart: Report of two cases. *Lancet*, 386, 1556–1563.

[43] Kim, E. J., Chen, C., Gologorsky, R., Santandreu, S. A., Torres, A., Wright, N., Goodin, M. S., Moyer, J., Chui, B. W., Blaha, C., Brakeman, P., Vartanian, S., Tang, Q., Humes, H. D., Fissel, W. H., and Roy, S. (2023). Feasibility of an implantable bioreactor for renal cell therapy using silicon nanopore membrane. *Nature Communications*, 14, Article 4890.

[44] Dasi, L. P., Simon, H. A., Sucosky, P., and Yoganathan, A. P. (2009). Fluid mechanics of artificial heart valves. *Clinical and Experimental Pharmacology and Physiology*, 36, 225–237.

[45] Min Yun, B., Dasi, L. P., Aidun, C. K., and Yoganathan, A. P. (2014). Computational modelling of flow through prosthetic heart valves using the entropic lattice-Boltzmann method. *Journal of Fluid Mechanics*, 743, 170–201.

[46] Kim, J. J., Kim, J., Hann, T., Kim, D., Roh, H. S., and Lee, S. J. (2019). Considerable drag reduction and fuel saving of a tractor–trailer using additive aerodynamic devices. *Journal of Wind Engineering and Industrial Aerodynamics*, 191, 54–62.

[47] Mannion, P., Toparlar, Y., Blocken, B., Clifford, E., Andrianne, T., and Hajdukiewicz, M. (2018). Aerodynamic drag in competitive tandem paracycling: Road race versus time-trial positions. *Journal of Wind Engineering and Industrial Aerodynamics*, 179, 92–101.

[48] Mitzokami, K., Kawakita, C., Kodan, Y., Takano, S., Higasa, S., and Shigenaga, R. (2010). Experimental study of air lubrication method and verification of effects on actual hull by means of sea trial. *Mitsubishi Heavy Industries Technical Review*, 47(3), 41–47.

[49] Huppert, H. E., and Neufeld, J. A. (2014). The fluid mechanics of carbon dioxide sequestration. *Annual Review of Fluid Mechanics*, 46, 255–272.

[50] Virk, P. S., Mickley, H. S. and Smith, K. A. (1970). The ultimate asymptote and mean flow structure in Toms' phenomenon. *Journal of Applied Mechanics*, 37, 488–493.

Index